Lecture Notes in Chemistry

Edited by G. Berthier, M. J. S. Dewar, H. Fischer
K. Fukui, H. Hartmann, H. H. Jaffé, J. Jortner
W. Kutzelnigg, K. Ruedenberg, E. Scrocco, W. Zeil

6

István Hargittai

Sulphone Molecular Structures

Conformation and Geometry from Electron Diffraction
and Microwave Spectroscopy; Structural Variations

Springer-Verlag
Berlin Heidelberg New York 1978

Author

István Hargittai
Central Research Institute of Chemistry
Hungarian Academy of Sciences
H-1525 Budapest, Pf. 17

Library of Congress Cataloging in Publication Data

Hargittai, István.
   Sulphone molecular structures.

   (Lecture notes in chemistry ; 6)
   Includes bibliographical references.
   1. Sulphones. 2. Molecular structure.
I. Title.
QD305.S6H37   547'.065   78-557

ISBN-13: 978-3-540-08654-3   e-ISBN-13: 978-3-642-93078-2
DOI: 10.1007/978-3-642-93078-2

2152/3140-543210·

Foreword

Recently, the molecular structures of a relatively large number of sulphone compounds have been elucidated in the vapour phase by electron diffraction and microwave spectroscopy. The main purpose of these studies is the determination of the sulphur bond configuration and the conformational properties. This leads to the observation and correlation of characteristic structural variations as various ligands are attached to the SO_2 group and as comparisons are made with related molecules.

Today it may be said that the structure of sulphone molecules is relatively well studied, and it appeared necessary to systematize the accumulated experimental data after critical considerations. This is done in the first part of this monograph. The second part presents the observed characteristic structural variations. Attempts are made to interpret these variations by valence shell electron pair repulsions and non-bonded interactions. Correlation relationships between geometric and vibrational parameters are also presented.

It is the metrical aspects of the molecular structure which are primarily considered. Since they correlate with other aspects of the molecular structure, e.g. electronic, it is hoped that the experimental information on the molecular geometry provides stimulus for further experimental, and, in particular, theoretical work on sulphones and related systems.

It is attempted to cover all electron diffraction and microwave spectroscopic investigations on sulphone molecules to date.
Admittedly, however, relatively larger weight is given to the
electron diffraction studies originating from the author's own
laboratory.

The core of this work was presented as an invited lecture to
the IV European Microwave Spectroscopy Conference organized under
the chairmanship of Professor Werner Zeil (Tübingen) as Conference
on Determination of Molecular Structure by Microwave Spectroscopy
and Electron Diffraction, Tübingen, March 21-25, 1977. It was
then Professor Zeil who suggested to put together this volume.

I would like to give here some words of appreciation.
Professor Lev V. Vilkov (Moscow) first turned my attention to the
structural problems of sulphones. Professor Sándor Lengyel made
invaluable contribution to creating the Budapest electron diffraction laboratory. Professor János Holló gave us support and
encouraged us at decisive moments in the studies of sulphur
stereochemistry.

Numerous colleagues and co-workers have participated in our
studies on sulphone structures. I have shared interest and work
in this field for years with Dr. Jon Brunvoll (Trondheim).
Especially significant contributions were made also by Dr. Magdolna
Hargittai, Mr. József Hernádi (deceased), Mrs. Mária Kolonits and

Dr. Ragnhild Seip (Oslo). This work also benefitted from the co-operation with Professor Sven J. and Dr. Bjørg N. Cyvins (Trondheim). I also appreciate the contributions from Dr. András Baranyi, Dr. K.P. Rajappan Nair (India), Dr. Béla Rozsondai, Dr. Ann Schmiedekamp (U.S.A.), Dr. György Schultz and Dr. Erzsébet Vajda.

The kindness of Dipl.-Chem. Petra Bulcke (Leipzig), Dr. Kolbjørn Hagen (Trondheim), Professor Kenneth Hedberg (Corvallis, Oregon), Dr. Bertram Nagel (Leipzig) and Dr. Germer Robinet (Toulouse) is gratefully acknowledged for communicating their results prior to publication. I express my sincere thanks to Dr. Barbara Mez-Starck (B.R.D.) for many useful references.

Typing this monograph was a special task performed by Mrs. Julia Simon with care and patience. Most of the Figures were drawn by Mrs. Judit Szilágyi.

Budapest, May 23, 1977

CONTENTS   page

Part One: MOLECULAR GEOMETRIES

Experimental determination of the molecular geometries

The metrical aspects of the sulphone molecules are dealt
with in the present discussion. They are only one, though a
very important one, of the characteristics of the molecular
structure. The metrical aspects of the molecule, i.e. its geometry, are determined by the relative spacial positions of the
atoms constituting the molecule. This is the same as to
characterize the molecule with all the internuclear distances,
or more descriptively, with the bond lengths, bond angles, and
the angles of internal rotation.

Only vapour-phase data are considered in detail since the
vapour phase is the only phase where the molecular structure
can be looked upon as determined solely by intramolecular
forces. This may be important in correlating the geometrical
parameters and bonding properties. Also, the conformational
behaviour may be strongly influenced by intermolecular interactions in the crystal or liquid phases and solutions.[*]

---

[*] Some examples of striking structural differences referring
to different phases are collected by Hargittai and Hargittai
(1977) and Hargittai and Paul (1977).

Besides, the crystal-phase molecular structure determinations
are carried out most often with X-ray diffraction which yields
distances between the centres of gravity of the electron
density distribution rather than internuclear distances. It is
also true, however, that many more sulphone structures have
been elucidated in the crystal phase than in the vapour phase.
Some of the results will be cited for comparison.

Let us turn now to the characterization and experimental
determination of the geometry of free molecules. In principle,
the unambiguous description of the molecular geometry would be
the equilibrium geometry, a hypothetically motionless structure,
which corresponds to the minimum of the potential energy function.
The equilibrium internuclear distances are called r_e parameters.
In reality, however, the molecules cannot be considered to be
rigid bodies, and the distance between the atoms is considerably influenced by the molecular vibrations and rotations.
This influence of the molecular vibrations and rotations
appears in a different way as different techniques are employed to determine the interatomic distances. In any study
where high accuracy is claimed and achieved, the vibrational
and rotational effects cannot be left out of consideration in
the interpretation of the results. On the other hand, it is to
be remembered that the results of the quantum chemical calculations refer to the equilibrium structure.

The two principal techniques for determining the molecular geometry in the vapour phase are electron diffraction
and rotational spectroscopy.

The electron diffraction method[*] is based on the phenomenon that a beam of fast electrons is scattered by the
potential resulting from the charge distribution in a molecule.
The resulting interference pattern depends on the molecular
geometry. The structure-dependent part of the electron
scattering intensity is determined not only by the interatomic
distances but also by the vibrational amplitudes. This
molecular intensity is proportional to the following expression:

$$M(s) \propto \exp\left(-\tfrac{1}{2}\,\ell^2 s^2\right)\,\sin[s(r_a - \kappa s^2)],$$

where s is an angular variable, r_a is an effective interatomic
distance, ℓ is the mean amplitude of vibration, and κ is the
so-called asymmetry parameter connected with anharmonicity.
The electron diffraction structure analysis can accordingly be
characterized as if it were the determination of the frequencies and the damping of the components of a sum of sine
functions. As the molecular intensity is Fourier-transformed,
another important but more descriptive function is attained,
which bears, somewhat unfortunately, the name of radial
distribution function, and which is indeed connected with the
probability distribution function of the intramolecular internuclear distances. It is a Gaussian-like distribution that

---

[*] For further reading see e.g. the following reviews: Bartell
(1972), Bauer (1970), Hargittai (1974a), Karle (1973), Seip
(1973), Schäfer (1976) and the book of Vilkov et al. (1974).

corresponds to an atomic pair of the molecule on the radial
distribution function. Approximate values for both the internuclear distance and the amplitude of vibration can be
directly read off this function. It can be shown that the r_a
distance parameter appearing in the expression for the
molecular intensity rigorously corresponds to the position of
the centre of gravity of the radial distribution function,
called $f(r)$ or $P(r)/r$. The $P(r)$ function itself is the
probability distribution function and $P_{ij}(r)dr$ is the probability that the distance between the ith and jth atoms has a
value in the interval of r and r + dr. The parameter corresponding to the position of the centre of gravity of the $P(r)$
function is called r_g and is the average value of the internuclear distance. The r_a effective internuclear distance
obtained directly from electron diffraction data, and the r_g
average internuclear distance, are connected with a good
approximation by the following relationship

$$r_g \approx r_a + \ell^2/r_a$$

It is important to emphasize here that electron diffraction
intensities are obtained as averaged from the molecules
distributed among the vibrational states. Accordingly, the
internuclear distances corresponding to the molecules in
excited states contribute to the average internuclear distance
with their proper weight, of course.

The average internuclear distance can be expressed in

terms of the equilibrium internuclear distance[*] in the
following way:

$$r_g = r_e + \delta r + \langle \Delta z \rangle + \frac{\langle \Delta x^2 \rangle + \langle \Delta y^2 \rangle}{2\, r_e} + \dots$$

This relationship refers to a Cartesian coordinate system
whose z-axis coincides with the equilibrium internuclear axis
and whose origin is the equilibrium position of one of the two
atoms. The term δr is the centrifugal distortion, or rather,
stretching, and Δx, Δy and Δz are the differences of the
displacements of the atoms in the directions of the three axes.
The quantity $\langle \Delta z \rangle$ vanishes in case of harmonic vibrations,
since it appears as a consequence of the anharmonicity of the
vibrational motion. The mean square perpendicular amplitudes
of vibration $\langle \Delta x^2 \rangle$ and $\langle \Delta y^2 \rangle$, however, are finite even
if there are only harmonic vibrations.

The average internuclear distance is not the same as the
distance between the average positions of the atomic nuclei

$$r_\alpha = r_e + \langle \Delta z \rangle ,$$

which differs from the equilibrium internuclear distance
because of anharmonicity only. The r_α parameter corresponds
to the thermal equilibrium at a given temperature, and for the
ground vibrational state

$$r_\alpha^o = r_e + \langle \Delta z \rangle$$

---

[*] On the various types of distance parameters see Kuchitsu and
Cyvin (1972) and references therein.

The distance between the average positions of the atomic nuclei can be obtained from the average internuclear distances by applying harmonic corrections.

The distance parameter r_g is the most convenient for characterizing the average length of a chemical bond (Kuchitsu, 1968). It has no such descriptive meaning, however, for distances between non-bonded atoms because of the effects of perpendicular vibrations (cf. Bastiansen-Morino shrinkage effect, see references in Cyvin, 1968). One of the merits of the r_α^o structure is that it is free from the effects of the perpendicular vibrations. Its other important merit is that this type of internuclear distance parameter provides us with a possibility to compare the electron diffraction results with the spectroscopic results on a common and well-defined physical basis.

The rotational spectroscopic determination of the molecular geometry is based on the following phenomenon. As energy is absorbed by the molecule, there are dipole transitions between quantized rotational states of the molecule. In the microwave[*] region, these rotational transitions happen without a change of the vibrational state; thus, the pure rotational spectrum can be determined. The rotational constants are

---

[*] Some books on molecular structure determination by microwave spectroscopy: Gordy and Cook (1970); Sugden and Kenney (1965); Wollrab (1967).

obtained from the frequencies of the rotational transitions. Since the rotational constants are related to the principal molecular moments of inertia, the molecular geometry can be deduced.

Since the rotational transitions corresponding to different vibrational states appear separately in the rotational spectrum, the internuclear distances corresponding to different vibrational states can be determined. Furthermore, with an extrapolation, for the simplest molecules it is possible to determine the equilibrium internuclear distance.

Up to three rotational constants can be obtained from a given isotopic species yielding three independent measurements. The molecular geometry of molecules larger than the simplest is determined, however, by a larger number of independent parameters. In such a case, further data from the rotational spectra of isotopically substituted species may be utilized. This is possible because the equilibrium internuclear distances are unchanged by the isotopic substitution, while there is usually an appreciable change in the atomic masses, that is, in the moments of inertia. The procedure, however, is complicated by the fact that the structure, even in the ground vibrational state, is somewhat different from the equilibrium structure because of the zero-point vibrations. Since the molecular vibrations are also mass-dependent, the influence of the zero-point vibrations will be somewhat different for various isotopically substituted species.

The internuclear distance which is obtained from the

rotational spectrum, or rather from the effective rotational constants as isotopic substitution is utilized, is an effective parameter and is called r_o (it usually corresponds to the ground vibrational state). This effective r_o parameter will not be invariant to the particular choice of isotopic substitutions performed. The difficulties can be demonstrated well by cases where it is possible to investigate more than the minimum number of isotopically substituted species. The values of the r_o parameters may differ even by hundredths of an Å from each other.

If the isotopic substitution is consequently and consistently performed for each atom of the molecule, and the so-called substitution coordinates of each atom obtained, it is possible to calculate the r_s substitution structure. This approximates well the equilibrium structure though it has no well-defined physical meaning.

It is possible to produce the parameters corresponding to the distances between average nuclear positions from the rotational spectroscopic data as well. The complicated correction procedures cannot be applied to the internuclear distance parameters though, but only to the rotational constants. The average structures, with well-defined physical meaning, can be obtained by considering the relationships between the effective and equilibrium values of the rotational constants

$A_o = A_e$ + anharmonic corrections + harmonic corrections

Employing the harmonic corrections, the average rotational

constants can be obtained

$$A_z = A_o - \text{harmonic corrections}$$

and from these the distance parameters corresponding to the average nuclear positions can be produced. These so-called r_z internuclear distances, derived from the microwave spectroscopy data, are essentially identical with the r_α^o internuclear distances attainable from the electron diffraction analysis. Accordingly, the comparison of the results in terms of this type of the internuclear distance parameters is the best way to compare the structural information obtained by the two techniques. Careful studies and correction procedures show that such average structures originating from electron diffraction and rotational spectroscopy are indeed not different.

The question of representation of the molecular geometry is especially important for molecules undergoing large amplitude motions. In such cases important features of the molecular geometry may be concealed by the effects of molecular vibrations.

Fortunately, so far at least as the sulphur bond configuration is concerned, the sulphone structures are relatively rigid systems and the various types of interatomic distances are not expected to differ considerably as a consequence of intramolecular motion. Thus e.g. calculated $K = \langle(\Delta z)^2 + (\Delta y)^2\rangle/2r$ correction terms may give some idea about the effects of the perpendicular vibrations on a similar system, $SOCl_2$ (Cyvin and Hargittai, 1968). The calculations gave $K(S=O) = 0.0014$ Å, $K(S-Cl) = 0.0004$ Å, $K(Cl...O) = 0.0002$ Å, and $K(Cl...Cl) = 0.0001$ Å at the temperature of 323 K. However, when subtle variations in the geometrical parameters are to be interpreted, the differences in the physical meaning of the internuclear distances originating from different measurements cannot be ignored.

Let us briefly review now the most important possibilities, limitations and sources of uncertainty for the determination of the sulphone molecular geometries in the vapour phase.

A prerequisite for an investigation is, of course, that it be possible to obtain sufficient vapour pressure of the compound without decomposition. In order to minimize the effects of the molecular vibrations, it is advantageous to perform the experiments at the lowest possible temperatures.

Microwave spectroscopy is limited to polar molecules. No analogous limitation exists for electron diffraction, on the contrary, the structural elucidation is greatly facilitated by high symmetry.

The frequencies of rotational transitions can be determined from the microwave spectrum with great precision and reproducibility. This is illustrated in Table 1 by some frequencies of rotational transitions of dimethyl sulphone as obtained in two independent studies. At the same time, it is also true that microwave spectroscopy is capable to determine only relatively simple structures. Concerning the sulphone molecules, the isotopic species occuring in their natural abundances do not provide sufficient data for a complete structural determination even of the simplest molecules. For improving this situation one possibility is to synthesize and investigate further isotopic

Table 1

Frequencies of rotational transitions of dimethyl sulphone from two
independent microwave spectroscopic studies

Transition		Jacob and Lide (1971)	Saito and Makino (1972)
2_{02}	3_{13}	25475.62	25475.84
2_{11}	3_{22}	26445.99	26445.73
2_{12}	3_{21}	26922.25	26922.32
2_{20}	3_{31}	27416.17	27415.78
2_{21}	3_{30}	27445.47	27445.40
3_{22}	4_{13}	33194.16	33194.18
3_{13}	4_{04}	33529.85	33529.75
3_{03}	4_{14}	33764.94	33765.00
3_{12}	4_{23}	34738.83	34738.66
3_{13}	4_{22}	35772.55	35772.55
3_{21}	4_{32}	35835.22	35835.07
3_{22}	4_{31}	35981.27	35981.21
3_{30}	4_{41}	36705.80	36705.33
3_{31}	4_{40}	36709.75	36709.35

species, which is not always easy and some times impossible. Another
possibility is to assume some of the parameters on the basis of
other data and then to determine the rest. The influence of such
assumptions has to be critically analysed in each case.

There is, however, a very fortunate circumstance which makes it possible to gain a limited amount of very reliable structural information on the sulphone molecules from their microwave spectra. The planar moment (or second moment) of the XSO_2Y molecules is determined by the masses and positions of the oxygen atoms situated outside the X-S-Y plane provided that the vibrational contribution (δ) due to the intramolecular motion can be ignored. The planar moment can be expressed the following way

$$I_a + I_b - I_c = m_O\ r^2(O...O) - \delta = \Delta \ ,$$

where I_a, I_b, and I_c are the principal moments of inertia, m_O is the mass of the oxygen atom, and $r(O...O)$ the distance between the two oxygen atoms. The above relationship is valid for molecules in which X and Y are, for example, halogen atoms. In case of dimethyl sulphone, however, only one of the hydrogen atoms in each methyl group is in the plane, while the other two also contribute to the planar moment:

$$I_a + I_b - I_c = m_O\ r^2(O...O) + 2\ m_H\ r^2(H...H) - \delta = \Delta \ .$$

Note that the planar moment is obtained from the principal moments of inertia not always by the above combination, the latter is determined by the orientation of the principal axes.

The planar moments of some of the sulphone molecules studied by microwave spectroscopy are collected in Table 2. The contributions from the methyl and methylene groups were estimated uniformly to be 3.266 amuÅ^2 and 3.730 amuÅ^2, respectively, using the values of 1.80 Å and 1.85 Å for the distances between the

Table 2

Planar moments in sulphone molecules investigated by microwave spectroscopy

Compound		Planar moments	
		Measured[*]	Corrected for hydrogen atoms[**]
FSO_2F	(i)	98.117	98.117
$ClSO_2Cl$	(ii)	98.753	98.753
FSO_2Cl	(iii)	98.700	98.700
FSO_2Br	(iv)	98.851	98.851
CH_3SO_2F	(v)	101.524	98.258
CH_3SO_2Cl	(vi)	101.904	98.638
CH_3OSO_2F	(vii)	102.009	98.743
$CH_3SO_2CH_3$	(v)	105.628	99.096
	(viii)	105.628	99.096
$\begin{smallmatrix} H_2C \\ | \quad SO_2 \\ H_2C \end{smallmatrix}$	(ix)	107.507	100.047
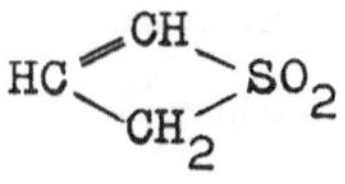 $\begin{smallmatrix} \quad CH \\ HC \quad SO_2 \\ \quad CH_2 \end{smallmatrix}$	(x)	101.547	97.817

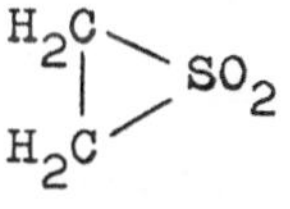

[*] $I_a + I_b - I_c$, except $\begin{smallmatrix} H_2C \\ | \quad SO_2 \\ H_2C \end{smallmatrix}$ and $\begin{smallmatrix} \quad CH \\ HC \quad SO_2 \\ \quad CH_2 \end{smallmatrix}$, for which

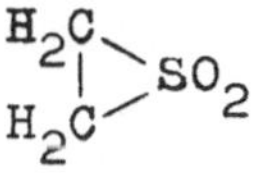

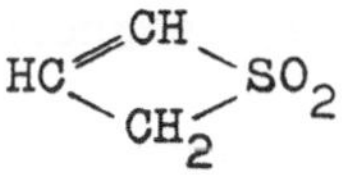

$I_a + I_c - I_b$, amu$Å^2$

** (v), (vi), (vii), (viii) I_{CH_3} = 3.266 amu$Å^2$ subtracted

for methyl group assuming r(H...H) = 1.80 Å;

(ix), (x) I_{CH_2} = 3.730 amu$Å^2$ subtracted for methylene

group assuming r(H...H)=1.85 Å

(i) Lide, Mann and Fristrom (1957); (ii) Dubrulle and Boucher
(1974); (iii) Holt and Gerry (1971); (iv) Raley, Wollrab and
Lovejoy (1973); (v) Jacob and Lide (1971); (vi) Van Eijck,
Korthoff and Mijlhoff (1975); (vii) Hargittai, Seip, Nair,
Britt, Boggs and Cyvin (1977); (viii) Saito and Makino (1972);
(ix) Nakano, Saito and Morino (1970); (x) Ralowski, Ljunggren
and Mjöberg (1973).

methyl hydrogens and methylene hydrogens in this order. The
corrected planar moments were obtained by subtracting the methyl
or methylene contributions from the planar moments calculated
from the principal moments of inertia. The corrected planar
moments appear to be very similar[*], indicating that the distances
between the oxygen atoms are very much the same in the various
sulphone molecules. Later, in the discussion of the structural
variations, we shall return to this remarkable constancy of the
oxygen...oxygen distances. Here it is noted only that this

---

[*] Their average value is 98.70 (σ = 0.58) amu$Å^2$, or omitting
the ring compounds, 98.64 (σ = 0.30) amu$Å^2$.

observation may be useful in facilitating the assignment of the rotational spectra as was the case indeed for fluorosulphuric acid methyl ester, for example (Hargittai, Seip, Nair, Britt, Boggs and Cyvin, 1977).

In estimating the uncertainty in the determination of the O...O distance from the planar moment, primarily the following two sources of error have to be considered: 1) the uncertainty in the assumption for the hydrogen...hydrogen distance, and 2) the possible deviation from zero of the δ vibrational correction term. An error of 0.05 Å in r(H...H) corresponds to an error of 0.002 Å in r(O...O). This is then increased to 0.003 Å if an uncertainty of ± 0.2 amuÅ^2 in the δ correction term is considered. The experimental error of the rotational constants have no appreciable effect in the determination of the O...O distance.

Electron diffraction may be successfully employed for the elucidation of more complicated structures. This is why the major part of the information on the conformational properties of sulphone molecules originates from electron diffraction. It is also true, however, that the electron diffraction data are more subject to misinterpretation than rotational spectroscopic data owing to the possibility of reproducing the experimental data equally well with various models.

The lengths of the S=O bonds as well as other bonds, particularly S-C and S-Cl can usually be determined very accurately from electron diffraction. There is, however, a special difficulty in the determination of the sulphur bond angles in sulphone molecules. The sulphur atom in sulphones has a tetrahedral configuration and, accordingly, there are very similar non-bond distances around the sulphur atom, thus they are very strongly correlated as regards the electron scattering data. Since the bond angles are calculated from the experimentally determined non-bond (and bond) distances, the strong correlation between the non-bond distances, and naturally, the mean vibrational amplitudes associated with them is the main hindrance to the unambiguous determination of the bond angles.

Of the sulphone molecules so far investigated, only sulphuryl chloride has a radial distribution (Figure 1) which

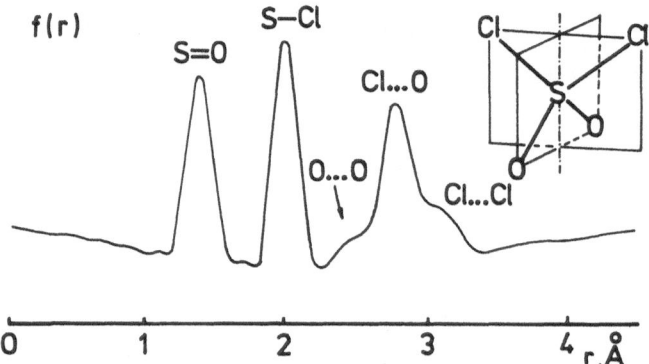

Figure 1

The radial distribution of sulphuryl chloride

shows distinguishable features for all different internuclear distances (bonds as well as non-bonds). On the radial distribution of the more compact sulphuryl fluoride (Figure 2) the

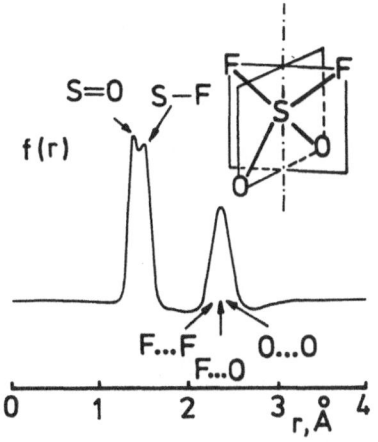

Figure 2

The radial distribution of

sulphuryl fluoride (after

Hagen, Coussens and Hedberg,

1975)

contributions from all non-bond interactions O...O, F...O, and
F...F, appear under the same maximum. The picture is more
complicated on the radial distribution of dimethyl sulphone
possessing the same C_{2v} symmetry as the previously mentioned
sulphuryl chloride and sulphuryl fluoride. The S...H distances
in dimethyl sulphone appear together with the non-bond distances
around the sulphur atom determining the sulphur bond angles
(Figure 3). Because of lower symmetry (C_s), there are more
interaction types at around the same place on the radial distributions of methane sulphonyl fluoride and methane sulphonyl
chloride (Figure 3). Here it is seen again that the non-bond
distances are more closely packed on the radial distribution of
the more compact methane sulphonyl fluoride than on that of
methane sulphonyl chloride.

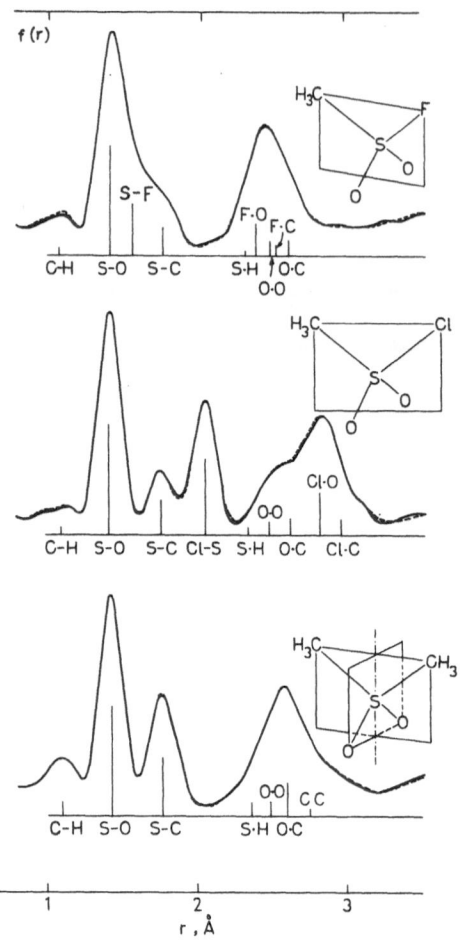

Figure 3

The radial distributions of methane sulphonyl fluoride, methane sulphonyl chloride and dimethyl sulphone

When there is strong correlation among the parameters, it may easily happen that the experimental data are equally well reproduced by more than one model and a decision among the models can be made only on the basis of further independent evidence. Relevant examples are given in the course of the discussion of individual structures.

The difficulties encountered in the electron diffraction determination of the non-bond distances around the sulphur atom on one hand, and the possibility of the unambiguous determination of the O...O distance from the microwave spectra on the other hand, led us to recognize (Hargittai and Hargittai, 1973a) the advantages of combining electron diffraction and microwave spectroscopic information in the determination of sulphone molecular geometries. This approach has been successfully applied in several instances (Hargittai and Hargittai, 1973a; Hargittai and Hargittai, 1974; Hargittai, Seip, Nair, Britt, Boggs and Cyvin, 1977).

We had to be aware of the problem of the difference in the physical meaning of parameters from the two techniques (electron diffraction r_a distances and microwave spectroscopic r_o O...O distance), since an uncritical combination of data could lead to introducing systematic errors. As was indicated already, however, the vibrational effects are supposed to be relatively small since the molecules in question, and especially the SO_2 fragments are quite rigid. It is suggested, however, to eliminate this uncertainty, if possible, since when the least-squares procedure converges, the constraint on the O...O distance can, in most cases, be removed, and a consistent set of parameters obtained.

Needless to emphasize that because of the strong correlation of the non-bond distances and the association with them

mean amplitudes of vibration, calculated amplitude values from
spectroscopic data provided great advantage in the electron
diffraction structure analyses.

The determined structures

The formulae of the sulphone molecules whose molecular
geometry has been determined by electron diffraction or microwave spectroscopy, are collected in Figure 4. There are more
electron diffraction than microwave spectroscopic studies to date.
Microwave spectroscopy has been employed for the simpler systems,
and predominantly for fluorides. The chlorides and the symmetri-
cally substituted derivatives are the largest and second largest
series, and were studied mainly by electron diffraction.

Sulphuryl fluoride, SO_2F_2 . It has been studied both by
microwave spectroscopy (Lide, Mann and Fristrom, 1957) and
electron diffraction (Hagen, Coussens and Hedberg, 1975). The
bond lengths and bond angles are collected in Table 3. Considering the experimental errors, the two data sets are not in
serious disagreement except, perhaps, the S=O bond length. The
discrepancies occuring in the S=O bond lengths and O=S=O bond
angles will be discussed in the sections concerning the
characteristic geometrical variations and the correlation

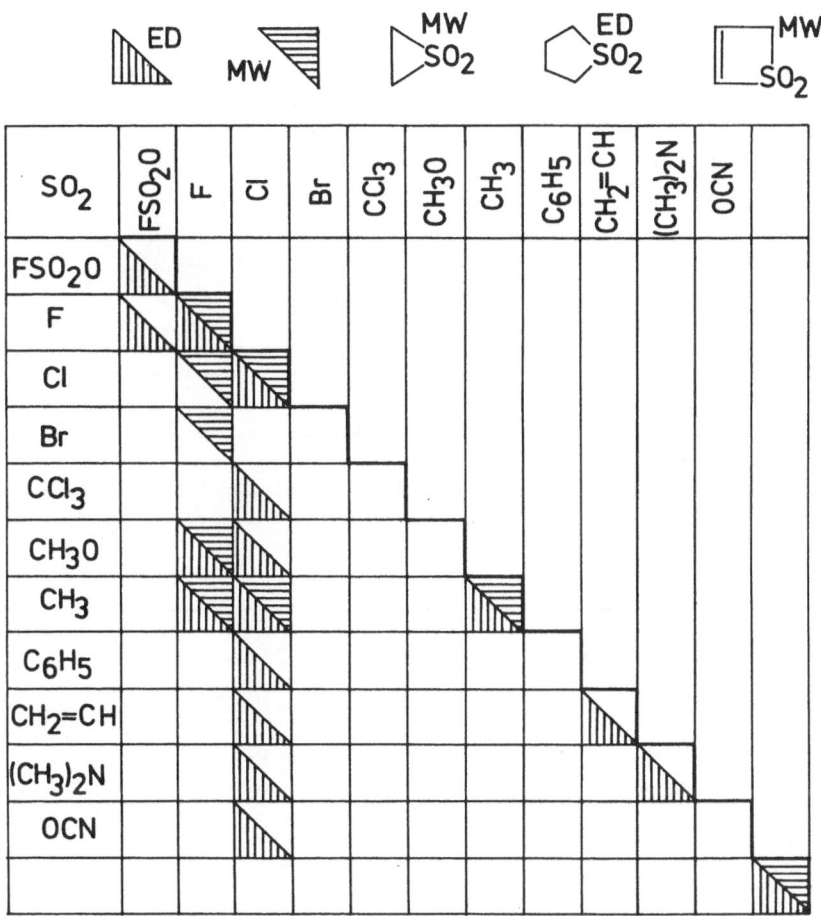

Figure 4

Sulphones, whose molecular geometry has been determined in the vapour phase

ED = electron diffraction;

MW = microwave spectroscopy

Table 3

Bond lengths (Å) and bond angles ($^{\circ}$) in sulphuryl fluoride
and sulphuryl chloride

X = F, Cl	SO_2F_2		SO_2Cl_2
	(a)	(b)	(c)
r(S=O)	1.405±0.003	1.397±0.002	1.404±0.004
r(S–X)	1.530±0.003	1.530±0.002	2.011±0.004
∠ O=S=O	124.0±0.2	122.6±1.2	123.5±1.0
∠ X–S–X	96.1±0.2	96.7±1.1	100.0±0.7
∠ X–S=O		108.6±0.2	107.7±0.4

(a) microwave spectroscopy, r_o parameters; Lide, Mann and
    Fristrom (1957)

(b) and (c) electron diffraction, r_a parameters

(b) Hagen, Coussens and Hedberg (1975)

(c) Hargittai (1969)

between the geometrical and vibrational parameters.

Sulphuryl chloride, SO_2Cl_2 . The bond lengths and bond
angles originating from electron diffraction (Hargittai, 1969)
are given in Table 3. These parameters were obtained in an
analysis employing complex electron scattering factors as in

all of our subsequent studies. This was preceded, however, by an analysis based on the same experimental data using atomic numbers as scattering factors. The results were relatively insensitive to the approximation used, but the changes in the empirical background showed noticeable shift in the S=O bond length (1.409 Å in the earlier study, Hargittai, 1968). There have also been microwave spectroscopic studies (Dubrulle and Boucher, 1974; Abbar, 1963; 1965; Abbar, Journel and Moise, 1965) which provided rotational constants only and no geometrical data were elucidated. Using the rotational constants communicated for ground state $SO_2{}^{35}Cl^{37}Cl$ (Dubrulle and Boucher, 1974):

$$A = 3459.39 \pm 0.02 \text{ MHz}$$
$$B = 2293.83 \pm 0.02 \text{ MHz}$$
$$C = 1888.16 \pm 0.04 \text{ MHz},$$

we calculated the O...O distance to be 2.485 Å.

Sulphuryl chloride fluoride, SO_2ClF . The microwave rotational spectra (Holt and Gerry, 1971) of the two most abundant isotopic species allowed an unambiguous determination of the O...O distance only (2.484 Å). Supposing that r(S=O) in sulphonyl chloride fluoride is intermediate (viz. 1.408 ± 0.006 Å) between those in sulphuryl fluoride (Lide, Mann and Fristrom, 1957) and in sulphuryl chloride (Hargittai, 1968), and that the F-S-Cl bond angle is intermediate (viz. $99 \pm 3°$) between ∠ F-S-F of SO_2F_2 and ∠ Cl-S-Cl of SO_2Cl_2 , the following bond lengths and bond angles were determined (Holt and Gerry, 1971):

$$r(S-F) = 1.55\pm0.02 \text{ Å}$$
$$r(S-Cl) = 1.985\pm0.015 \text{ Å}$$
$$\angle\ O{=}S{=}O = 123.7\pm1^{\circ}$$
$$\angle\ O{=}S{-}F = 107.5\pm2^{\circ}$$
$$\angle\ O{=}S{-}Cl = 107.5\pm2.5^{\circ}\ .$$

Sulphuryl bromide fluoride, SO_2BrF . In a microwave spectroscopic study (Raley, Wollrab and Lovejoy, 1973) similar to the one on SO_2ClF, using the following assumptions $r(S{=}O)$ = 1.407 Å, $r(S-F)$ = 1.56 Å, and $\angle\ O{=}S{=}O$ = 123.7°, the following two parameters were determined:

$$r(S-Br) = 2.155\pm0.006 \text{ Å}$$
$$\angle\ F{-}S{-}Br = 100.6\pm1.5^{\circ}$$

The mean vibrational amplitudes of simple sulphuryl halides as determined from electron diffraction and calculated from spectroscopic data are collected in Table 4.

Polysulphuryl fluorides, $S_2O_5F_2$ and $S_3O_8F_2$. Two groups of models have been tested in the electron diffraction study of $S_2O_5F_2$ (Hencher and Bauer, 1973). In one group of models both SO_2F groups are rotated by the same angle around the S-O bonds ensuring at least C_2 symmetry. In the other group of models the rotation of the two SO_2F groups were characterized by two different and independent angles. Hencher and Bauer (1973) noted that all combinations of rotation angles occuring in all possible staggered and eclipsed conformations have been used as starting parameters in the least-squares refinements. These possible forms are compiled in Table 5. The rotation angles are labelled consistently with the numbering in Figure 5. The

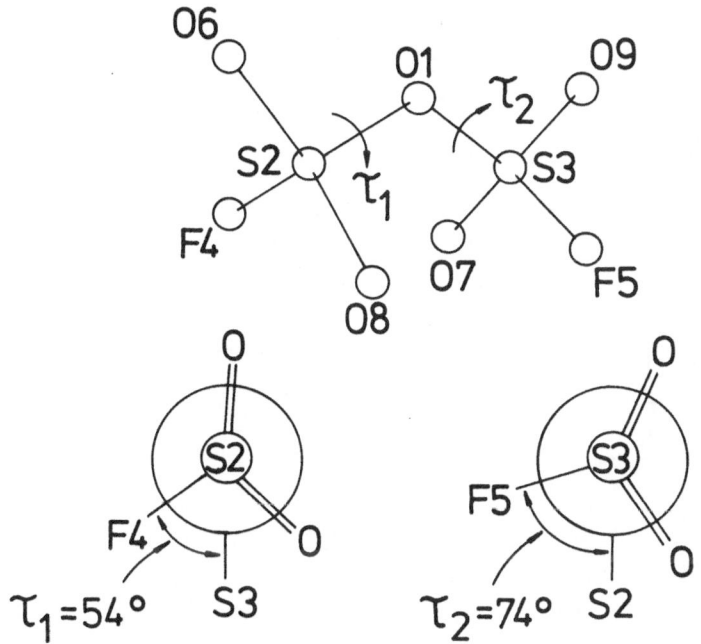

Figure 5

The molecular model of $S_2O_5F_2$, the characterization of the angles of rotation, and the Newman projections of the conformer established by Hencher and Bauer (1973)

rotation angles are 0° when the F-S-O-S-F chain is in one plane. The two rotation angles have the same sign when, looking from

Table 4
Mean amplitudes of vibration (Å) in sulphonyl halide molecules

	SO$_2$F$_2$		SO$_2$Cl$_2$		SO$_2$FCl	SO$_2$FBr
	SP**	ED**	SP**	ED**	SP**	SP**
S=O	0.034(i) 0.040(ii) 0.024(iii)	0.033(2)(iv)	0.035(v) 0.040(ii)	0.035(7)(v)	0.035(vi) 0.034(vii)	0.035(vi) 0.035(vii)
S–F	0.041(i) 0.043(ii) 0.024(iii)	0.040(2)(iv)			0.043(vi) 0.042(vii)	0.043(vi) 0.043(vii)
S–Cl			0.047(v) 0.051(ii)	0.046(4)(v)	0.044(vi) 0.049(vii)	
S–Br						0.048(vi) 0.049(vii)
O...O	0.052(i)* 0.060(i)** 0.058(ii) 0.056(iii)	0.071(7)(iv)	0.066(v) 0.061(ii)	[0.061](v)	0.056(vi) 0.058(vii)	0.055(vi) 0.056(vii)
F...F	0.057(i)* 0.060(i)** 0.066(ii) 0.081(iii)	0.071(7)(iv)				

Table 4 (continued)

F...Cl					0.106(vi) 0.104(vii)	
F...Br						0.084(vi) 0.081(vii)
Cl...Cl			0.099(v) 0.085(ii)	0.101(18)(v)		
O...F	0.064(i)* 0.060(i)** 0.058(ii) 0.069(iii)	0.071(7)(iv)			0.062(vi) 0.060(vii)	0.063(vi) 0.059(vii)
O...Cl			0.069(v) 0.064(ii)	0.072(6)(v)	0.069(vi) 0.065(vii)	
O...Br						0.075(vi) 0.073(vii)

* Temperature 298 K; **SP - spectroscopic calculations, ED - electron diffraction

(i) Cyvin and Hargittai (1969); (ii) Venkateswarlu and Malathi Devi (1965); (iii) Pichai, Krishna Pillai and Ramaswamy (1967); (iv) Hagen, Coussens and Hedberg (1975); (v) Hargittai and Cyvin (1969); (vi) Cyvin and Cyvin (1972); (vii) Ramaswamy and Jayaraman (1971)

Table 5

The characterization of possible rotamers of $S_2O_5F_2$[*]

Rotation angles		Conformations[**]		Symmetry
τ_1(F4–S2–O1–S3)	τ_2(F5–S3–O1–S2)	1(τ_1)	2(τ_2)	
0	0	e	e	C_{2v}
0	120	e	e	C_1
120	120	e	e	C_2
120	240	e	e	C_s
0	60	e	s	C_1
0	180	e	s	C_1
60	60	s	s	C_2
60	180	s	s	C_1
60	300	s	s	C_s
180	180	s	s	C_{2v}

[*] After Hencher and Bauer (1973)

[**] e = eclipsed, s = staggered

the sulphur atom towards the bridging oxygen, the rotation around the S–O bond has the same direction. The least-squares refinements provided five rotation angle combinations with local minima of which the combination τ_1(F4–S2–O1–S3) = 53.8±2.7° and τ_2(F5–S3–O1–S2) = 73.7±2.4° was chosen on the basis of statistical criteria. The corresponding form is demonstrated by Figure 5. An important constraint was the assumption of only one conformer to be present in this structure analysis. This constraint and the large number of possible conformers make the conclusions as regards the conformational properties of $S_2O_5F_2$ less unambiguous. It is important to note at the same time that the geometrical parameters changed very little throughout the structure analysis when various conformations have been tested. Further uncertainty originates, of course, from the application of assumed mean amplitudes of vibration (l values). They all seem to be reasonable: $l(S=O) = 0.04$ Å, $l(S-O) = l(S-F) = 0.05$ Å, $l(O...O)$ [O–S=O] $= l(F...O)$ [F–S=O] $= l(O...O)$ [O=S=O] $=$ $= 0.06$ Å. The mean amplitudes of vibration associated with the four-member chains were assumed to be the same and 0.088 Å was determined for them. All the other l values were also assumed to be the same and 0.123 Å was obtained for them from the refinement. The bond lengths and bond angles are given in Table 6.

The difficulties mentioned in the general discussion in connection with the electron diffraction investigation of sulphone molecular geometries are of course relevant for the polysulphuryl fluorides as well. It is possible, however, to determine the bond angle at the bridging oxygen atom very accurately.

The conformational properties of the $S_3O_8F_2$ molecule (Figure 6) may be even more complicated than those of $S_2O_5F_2$. Thus even more constraint had to be applied in the structure analysis of $S_3O_8F_2$ (Hencher and Bauer, 1973). No details are given here but the geometrical parameters reported are reproduced in Table 6.

Table 6

Bond lengths (Å) and bond angles (°) in two polysulphuryl
fluorides (Hencher and Bauer, 1973)

	$S_2O_5F_2$	$S_3O_8F_2$
r_g(S=O)	1.398±0.002	1.402±0.003
r_g(S–F)	1.525±0.005	1.525±0.012
r_g(S–O)	1.611±0.005	1.613±0.006
∠ O–S–O		97.8±1
∠ S–O–S	123.6±0.5	123.6±1.2
∠ O–S=O	106.1±0.9	106.5±0.8
∠ F–S=O	106.6±0.6	105.5±1.2
∠ O–S–F	102.4±1.8	101.3±1.5
∠ O=S=O	126.8±1.2	128.5±1.4 (128.8±1.4)[*]

[*] Refers to the bridging SO_2 group

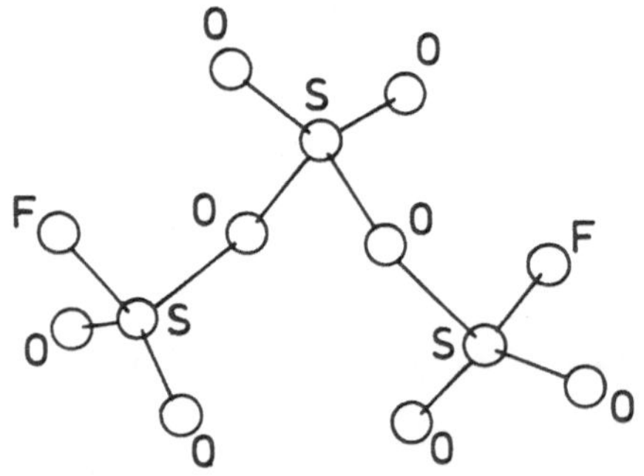

Figure 6

The molecular model of $S_3O_8F_2$
(after Hencher and Bauer, 1973)

Fluorosulphuric acid methyl ester, CH_3OSO_2F . The microwave
spectroscopic and electron diffraction investigations as well as
vibrational spectroscopic calculations have been performed con-
currently (Hargittai, Seip, Nair, Britt, Boggs and Cyvin, 1977).
One conformer possessing a symmetry plane was identified from
the microwave spectra. Considering the F–S–O–C chain (Figure 7)

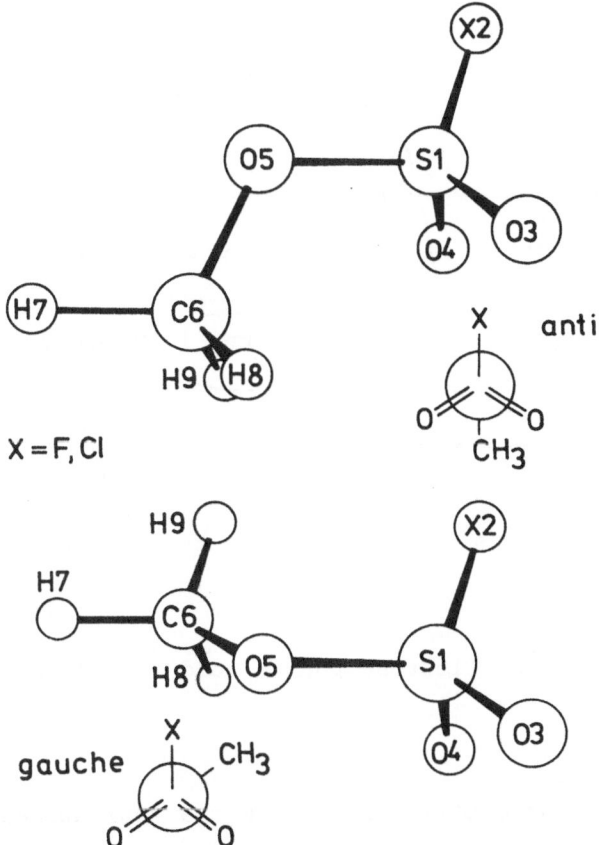

Figure 7
The molecular models and Newman projections of two conformers
of fluorosulphuric acid methyl ester and chlorosulphuric acid
methyl ester

this conformer could be the anti or the syn form. According to
the electron diffraction data this is unambiguously the anti form.
At the same time the electron diffraction data also indicated that
a considerable amount of gauche form may be present. However,
since the rotation-dependent contribution to the total scattering
is relatively small (cf. the radial distributions in Figure 8),

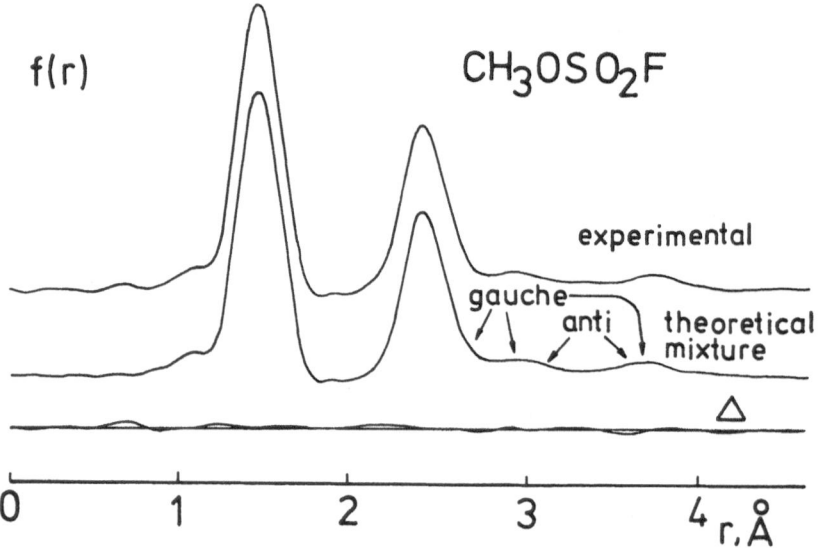

Figure 8

Radial distributions of fluorosulphuric acid methyl ester

and what is more, the contributions from the anti and gauche
forms are not easily distinguishable (due to the similarity of
scattering power from atomic pairs F...C and O...C), no definite
conclusion could be drawn. Under the conditions of the microwave

experiment no more than a minor amount of a gauche form could
have been present.

Minimal basis set (STO-6G) ab initio molecular orbital
calculations also indicated that the stable conformer of fluorosulphuric acid methyl ester is the anti form. A secondary minimum
was found at a gauche conformation with an energy higher by a
fraction of one kilocalorie. The syn form was seen to be less
favourable by an amount on the order of 10 kcal. Without giving
any importance to the numerical values, the general shape of the
potential curve is thought to be meaningful.

The geometrical parameters as determined in the electron
diffraction analysis are shown in Table 7. The results appeared
to be little sensitive to the changing conformational properties
of the models. It was of great advantage that the O...O distance
with great accuracy was available from the microwave spectra.
Besides, in addition to the requirement of convergency in the
least-squares refinement, an important criterion for any acceptable structure was the consistency with the rotational
constants derived from the microwave spectra. It is stressed,
however, that such a criterion was considered to a certain limit
only. Discrepancies up to 2% were acceptable since the parameters
were not corrected for the effects of the intramolecular motion,
and, what may be even more important, the standard deviations
were relatively large.

The mean amplitudes of vibration from spectroscopic calculations were very useful in the electron diffraction structure
analysis. They are given in Table 8. They are, however, only of

Table 7

Bond lengths (Å) and bond angles (°) in fluorosulphuric
acid methyl ester and chlorosulphuric acid methyl ester
from electron diffraction

X = F, Cl	CH_3OSO_2F (a)	CH_3OSO_2Cl (b)
$r_a(S=O)$	1.410±0.002	1.419±0.003
$r_a(S-X)$	1.545±0.006	2.023±0.004
$r_a(S-O)$	1.558±0.007	1.562±0.004
∠ O=S=O	124.4±0.7	122.2±1.5
∠ X-S=O	106.8±0.5	106.4±0.6
∠ O-S=O	109.5±0.6	108.7±0.8
∠ X-S-O	96.8±0.6	102.8±1.4
∠ S-O-C	116.5±0.7	114.4±1.1

(a) Hargittai, Seip, Nair, Britt, Boggs and Cyvin (1977)

(b) Hargittai, Schultz and Kolonits (1977)

tentative character since, lacking experimental vibrational
spectroscopic data, the calculations were based on assumptions
and force fields of other molecules, constructed again rather
tentatively, viz. fluorosulphuric acid (Cyvin and Hargittai,
1969) and methane sulphonyl fluoride (Cyvin, Dobos, Hargittai,
Hargittai and Augdahl, 1973).

Table 8

Calculated mean amplitudes of vibration (Å) for important
distances in fluorosulphuric acid methyl ester and chlorosulphuric acid methyl ester[*]

| X = F, Cl | CH_3OSO_2F | CH_3OSO_2Cl |
	(a)	(b)
Rotation independent		
S=O	0.035	0.035
S–X	0.042	0.043
S–O	0.049	0.049
O3...O4	0.056	0.056
O3...O5	0.065	0.074
O3...X	0.062	0.070
O5...X	0.097	0.105
S...C	0.064	0.064
Rotation-dependent		
X...C (anti)	0.088	0.094
X...C (gauche)	0.122	0.137
O3...C (anti)	0.092	0.100
O3...C (gauche)	0.094	0.106
O4...C (gauche)	0.068	0.072

[*] At 298 K, the numbering of atoms is given in Figure 7

(a) Hargittai, Seip, Nair, Britt, Boggs and Cyvin (1977)

(b) Cyvin (1975)

<u>Chlorosulphuric acid methyl ester</u>, CH_3OSO_2Cl . This molecule
is better suited for an electron diffraction determination
(Hargittai, Schultz and Kolonits, 1977) than its fluorine analog
since the contribution of the most important rotation-dependent
distances, Cl...C and O...C (cf. Figure 7), differ considerably.
On the other hand, attempts for microwave spectroscopic investigation have been unsuccessful (Boggs, 1975). The experimental
electron diffraction data (nozzle at about room temperature)
could be well approximated (Figure 9) by gauche conformer only,
in which the angle of rotation is around 74° (anti form corresponds
to 180 °).   On the basis of dipole moment measurements in
solution, Exner, Dembech and Vivarelli (1971) concluded that the
dihedral angle in CH_3OSO_2Cl may have values from 0 to 60°
(according to our reference system), and they excluded the anti
form. Considering the uncertainty in the electron diffraction
determination of the rotation angle and the inherent uncertainties
in the dipole moment calculations, the agreement is satisfactory.
Calculated dipole moments as a function of the rotation angle of
CH_3OSO_2Cl are shown in Figure 10.

Although the experimental radial distribution (Figure 9)
seemed to indicate the presence of the anti form, the least-
squares refinements showed that its amount is certainly less
than 20%.

Calculated mean amplitudes of vibration (Cyvin, 1975) shown
in Table 8 were utilized in the structure analysis of chlorosulphuric acid methyl ester. Because of the strong correlation
of the parameters, the O...O distance was fixed in most of the

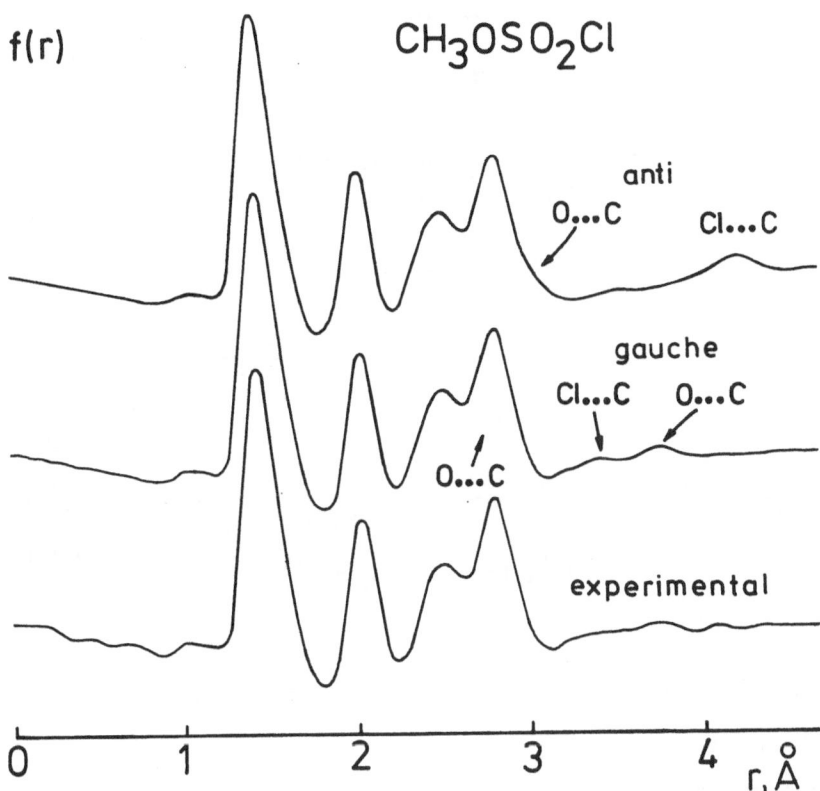

Figure 9

Radial distributions of chlorosulphuric acid methyl ester. The
positions of the most important rotation-dependent distances
are shown

refinements at a value (2.485 Å) consistent with the data on a
relatively large series of sulphones. The geometrical parameters
determined are given in Table 7. The uncertainties were obtained

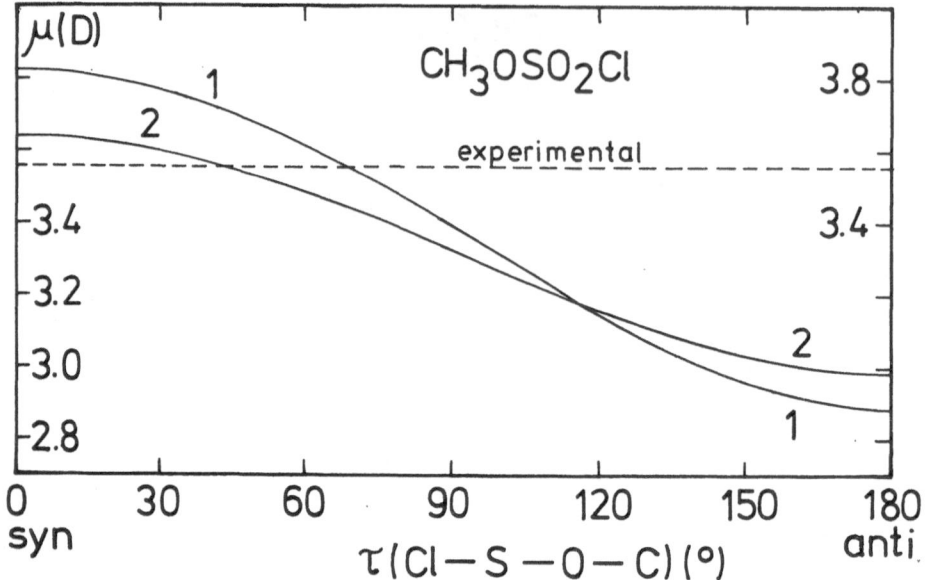

Figure 10

Experimental and calculated dipole moments of chlorosulphuric acid methyl ester

1 Bond moments and assumed geometry from Exner, Dembech and Vivarelli (1971)

2 Bond moments from Exner et al. (1971), geometry from electron diffraction

in the following way (this procedure was typical for similar analyses): the total error included the standard deviations from the least-squares refinement multiplied by $\sqrt{3}$ for data correlation and an estimated 0.2% experimental error for the distances. In addition, the errors were increased by adding the changes in the parameters that occured when the assumed values of $r(O...O)$ were changed as to cover the interval of 2.470 to 2.500 Å.

Methane sulphonyl fluoride, CH_3SO_2F . The staggered form relative to the rotation around the S-C bond is characterized by Figure 11. The barrier to rotation was found to be 2.52 ± 0.35

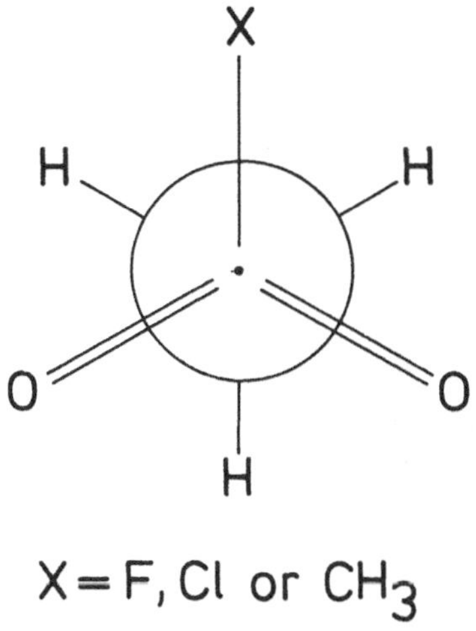

Figure 11
Newman projection of the molecular models of methane sulphonyl fluoride, methane sulphonyl chloride, and dimethyl sulphone representing view along the S-C bond

$X = F, Cl$ or CH_3

kcal mol^{-1} from microwave spectroscopic data (Jacob and Lide, 1971). The $O...O$ distance (2.480 Å) determined from the microwave spectra was assumed in the electron diffraction structure analysis (Hargittai and Hargittai, 1973a) which produced the geometrical data collected in Table 9. This was the first in the series of similar combined analyses. It was realized, of course that the microwave spectroscopic value for $r(O...O)$ was an r_o parameter while the electron diffraction least-squares refinement based on the molecular intensity yielded r_a parameters. For

Table 9

Bond lengths ($\AA$) and bond angles ($^{\circ}$) in methane sulphonyl fluoride and methane sulphonyl chloride from electron diffraction

X = F, Cl	CH_3SO_2F (a)	CH_3SO_2Cl (b)
r_a(S=O)	1.410±0.003	1.424±0.003
r_a(S-X)	1.561±0.004	2.046±0.004
r_a(S-C)	1.759±0.006	1.763±0.005
∠ O=S=O	123.1±1.5	120.8±0.8
∠ X-S=O	106.2±0.4	107.1±0.7
∠ C-S-X	98.2±1.5	101.0±1.4

(a) Hargittai and Hargittai (1973a)

(b) Hargittai and Hargittai (1973c)

estimating the errors, the refinements have been repeated with assuming r(O...O) = 2.47 and then also 2.49 $\AA$. A change of 0.01 $\AA$ in r(O...O) resulted in a change of 0.85° in the O=S=O bond angle. Of the other geometrical parameters, only the F-S-C bond angle showed appreciable change (1.1°). All this was taken onto account in producing the uncertainties given in Table 9.

Methane sulphonyl chloride, CH_3SO_2Cl . The electron diffraction analysis (Hargittai and Hargittai, 1973c) provided three models reproducing equally well the experimental data.

The experimental molecular intensities and difference curves for the three models are shown in Figure 12. The models differed in

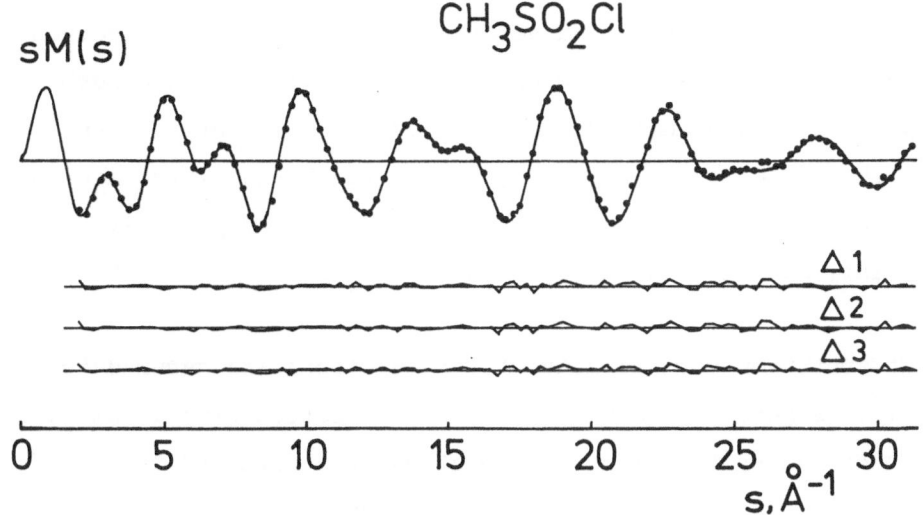

Figure 12

Experimental (dots) and theoretical (full line) molecular intensities of methane sulphonyl chloride. The difference curves refer to models differing mainly in the O=S=O bond angle (1, 120.8°, 2, 124.2°, and 3, 127.5°)

the O=S=O bond angle (120.8, 124.2, and 127.5°) and some of the mean amplitudes of vibration. On the basis of structural data accumulated on sulphone molecules by the time of the investigation and selecting also the most reasonable mean amplitudes of vibration, the model with O=S=O = 120.8° was preferred. Unambiguous solution could be provided, however, only by microwave spectroscopy. Such an investigation was kindly performed at our suggestion by van Eijck, Korthoff and Mijlhoff (1975) employing double resonance technique and the preferred model was confirmed. The geometrical parameters determined in the electron diffraction investigation are presented in Table 9. The uncertainty of the O=S=O angle was originally given as three times the estimated total error (2.4°), but in the light of the microwave spectroscopic evidence, we have no reason today to give the uncertainty for this parameter in any way different from the others.

The rotation form relative to the S-C bond was assumed to be staggered (Figure 11) in the electron diffraction study. This was also confirmed by the microwave spectroscopic data and the barrier to rotation was found to be at least 2 kcal mol^{-1}.

The mean amplitudes of vibration of methane sulphonyl fluoride and methane sulphonyl chloride as obtained from spectroscopic calculations and determined from the electron diffraction data are compiled in Table 10.

Trichloromethyl sulphonyl chloride, CCl_3SO_2Cl . Its molecular model is shown in Figure 13. The report on its electron diffraction investigation (Alekseev, 1967) contained no information as regards the mean amplitudes of vibration assumed or determined. They are thought to be important in determining the geometrical parameters of this relatively compact molecule. The following bond lengths and bond angles were reported: r(S=O) =

Table 10

Mean amplitudes of vibration (Å) in methane sulphonyl fluoride, methane sulphonyl chloride and dimethyl sulphone

	CH_3SO_2F		CH_3SO_2Cl		$CH_3SO_2CH_3$	
	SP(i)	ED(ii)	SP(i)	ED(iii)	SP(iv)	ED(v)
S=O	0.036	0.039(2)	0.037	0.043(2)	0.037	0.042(1)
S–F	0.048	0.052(3)				
S–Cl			0.061	0.057(2)		
S–C	0.047	0.057(4)	0.065	0.044(4)	0.044	0.057(1)
C–H	0.078	0.063(9)	0.078	0.076(13)	0.078	0.075(2)
O...O	0.069	[0.057]	0.095	0.066(26)	0.063	0.052(14)
C...F	0.069	0.056(24)				
C...Cl			0.079	0.133(50)		
C...C					0.062	0.058(21)
O...F	0.081	0.054(6)				
O...Cl			0.095	0.087(7)		
O...C	0.069	0.077(18)	0.083	0.078(24)	0.061	0.071(8)
S...H	0.106	0.149(43)	0.112	0.106(37)	0.105	0.120(10)
	0.112		0.120			

SP – spectroscopic calculations, ED – electron diffraction
(i) Cyvin, Dobos, Hargittai, Hargittai and Augdahl (1973), T = 105 °C; (ii) Hargittai and Hargittai (1973a), T = 105 °C; (iii) Hargittai and Hargittai (1973c), T = 110 °C; (iv) Cyvin (1973), T = 25 °C; (v) Hargittai and Hargittai (1974), T = 130 °C

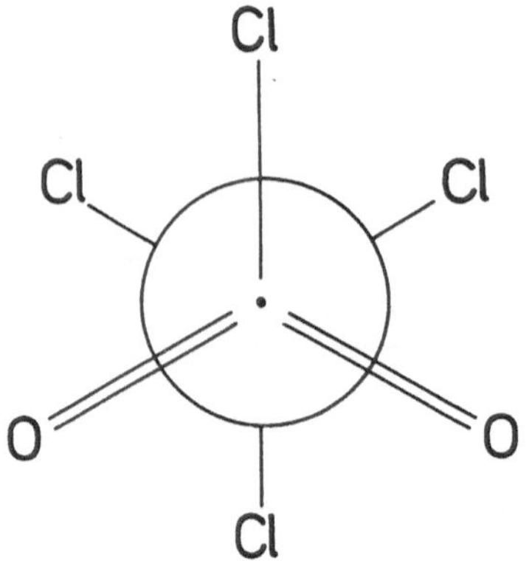

Figure 13

Newman projection of the molecular model of trichloromethyl sulphonyl chloride representing view along the S-C bond

= 1.45±0.01 Å, r(C-Cl) = 1.76±0.01 Å, r(S-C) = 1.81±0.02 Å, r(S-Cl) = 2.03±0.02 Å, ∠O=S=O = 111±2°, and ∠C-S-Cl 96±3°. The cited investigation concentrated on examining possible deviations of the Cl-C-Cl bond angles from the ideal tetrahedral and of the threefold symmetry axis of the trichloromethyl group from the S-C bond. In the light of the structural data for the relatively large series of sulphone molecules, the S=O bond length seems to be too large while the reported O=S=O bond angle is certainly felt to be too small. This important compound calls for a reinvestigation which should involve spectroscopic information.

Dimethyl sulphone, $(CH_3)_2SO_2$. Several investigations have been performed recently in order to elucidate its molecular structure. The orientation of the methyl group around the S-C

bond was such that the C-H bonds staggered the other sulphur
bonds (Figure 11). Both the semiempirical molecular orbital
calculations (Robinet, Crasnier, Labarre and Leibovici, 1972)
and experimental evidence (Clever and Westrum, 1970) gave support
for this model. The rotational barriers for one CH_3 group were
found to be 3.65 and 3.4 kcal mol^{-1}, in the cited studies,
respectively.

Table 11 presents the bond lengths and bond angles characterizing the sulphur bond configuration in dimethyl sulphone
from some of the recent studies. There persisted a controversy
for some time between the O=S=O bond angles determined by microwave spectroscopy (see Table 11) and by an earlier electron
diffraction study (Oberhammer and Zeil, 1970), the latter re-
porting $127.1\pm2.4^\circ$. As has been shown later for other simple
sulphones, however, electron diffraction alone may not be
sufficient to arrive at a unique solution of the structure.
Accordingly, we decided that an electron diffraction reinvestigation utilizing spectroscopic data was warranted (Hargittai and
Hargittai, 1974). The microwave spectroscopic datum on $r(O...O)$
has been used as a constraint at the first stage of our structure
analysis, but this constraint was removed in the final refinement
of the parameters. When convergency is achieved, and provided
that we are in the true minimum, this is perhaps the best
approach since it avoids introducing a systematic error by using
an r_0 parameter as constraint in the refinement of the other r_a
parameters. In estimating the total errors, the sensitivity of
the parameters to the value of $r(O...O)$ was also considered.

Table 11

Bond lengths ($\mathring{A}$) and bond angles ($^\circ$) of dimethyl sulphone
from vapour-phase (a, b, c) and crystal-phase (d, e)
determinations

	(a1)	(a2)	(b)	(c)
r(S=O)	1.433	[1.425]	1.431±0.003	1.435±0.003
r(S–C)	[1.770]	1.780	1.777±0.006	1.771±0.004
∠ O=S=O	120.8	122.0	121.0±0.3	119.7±1.1
∠ C–S–C	103.4	102.7	103.3±0.2	102.6±0.9

(a), (b) microwave spectroscopy, r_o parameters

(a) Jacob and Lide (1971); (b) Saito and Makino (1972)

(c) electron diffraction, r_a parameters, Hargittai and
Hargittai (1974)

	(d)	(e)
r(S=O)	1.445±0.016	1.446±0.003
r(S–C)	1.778±0.017	1.765±0.005
∠ O=S=O	117.9±0.8	117.3±0.2
∠ C–S–C	103.0±0.8	104.8±0.2

(d), (e) X-ray diffraction

(d) Sands (1963); (e) Langs, Silverton and Bright (1970)

This electron diffraction study provided a structure consistent
with the microwave spectroscopic results (Table 11). The findings
from X-ray diffraction studies on the crystal-phase structure are
also reproduced in Table 11. Spectroscopically calculated mean
amplitudes of vibration together with those determined from the
electron diffraction data are given in Table 10.

It is of interest to compare the microwave spectroscopic and
electron diffraction results in terms of rotational constants as
done in Table 12. Here again it is to be remembered that discrepancies up to 1 or 2% between the microwave spectroscopic and
electron diffraction data may be due to the difference in the
physical meaning of the parameters originating from the two techniques, and that the uncertainties of the electron diffraction
parameters render a more detailed comparison meaningless. Changes
of 0.005 Å and 0.5° in the bond lengths and bond angles may
correspond typically to about 20 and 35 MHz changes in the rotational constants, respectively. This is also why we did not
give the rotational constants calculated from our structural data
with more digits. In the light of the above warnings, the agreement between our electron diffraction data (column c, Table 12)
and the microwave spectroscopic data is completely satisfactory.
At the same time, the C rotational constant obtained from the
results of the earlier electron diffraction study (Oberhammer and
Zeil, 1970) is certainly too large. Since this rotational
constant is determined primarily by the positions of the oxygen
atoms, the correlation with the reported too large $O=S=O$ bond
angle is obvious.

Table 12

Rotational constants (MHz) measured and calculated for
dimethyl sulphone

	(a)	(b)	(al)	(c)	(x)
A[*]	4638.202±0.006	4638.15±0.02	4639.92	4595	4670.51
B[*]	4295.405±0.006	4295.44±0.02	4295.49	4338	4239.12
C[*]	4177.107±0.007	4177.13±0.02	4178.44	4171	4317.28

[*] A, axis perpendicular to the OSO plane; B, C_2 axis; C,
   axis perpendicular to the CSC plane
(a) and (b) experimental data; (al) and (c) models,
references in Table 11; (x) calculated by Jacob and Lide
(1971) from parameters reported by Oberhammer and Zeil
(1970), see text

We have compiled a test to examine how is it possible to
arrive at an erroneous answer in similar situations. For a series
of dimethyl sulphone structure refinements the generalized
R factor

$$R = \left(\frac{\sum_{s=2.50}^{34.25} [M^E(s) - M^T(s)]^2 W_s}{\sum_{s=2.50}^{34.25} [M^E(s)]^2 W_s} \right)^{1/2}$$

has been computed by keeping the O=S=O angle at fixed values and refining all the other parameters. The results are shown in Figure 14. The minima "c" and "f" correspond to schemes, where

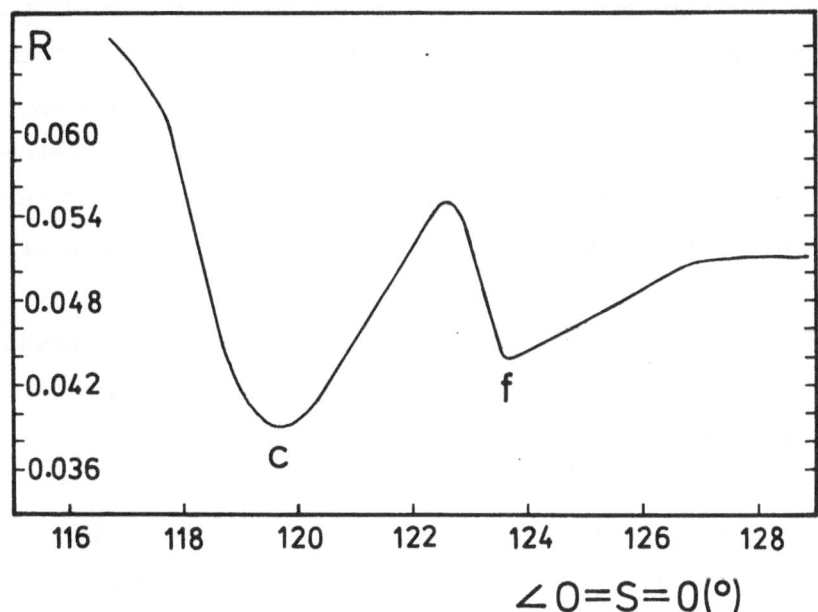

Figure 14

Generalized R factor as plotted against the O=S=O bond angle values fixed in separate refinements. The minima c and f correspond to refinement schemes in which this bond angle was also refined

the O=S=O angle was also refined in addition to the other parameters. The minimum denoted as "c" (and corresponding to the results given in columns c of Tables 11 and 12) is considered to be the correct model for $(CH_3)_2SO_2$ not only and not primarily because it is associated with a smaller R value than the other minimum denoted as "f". In addition to being a minimum, the structure corresponding to this minimum "c" is the one consistent with the microwave spectroscopic data. The other minimum "f" in our calculations, which corresponds to a structure containing $\angle O{=}S{=}O = 123.7°$ and $r(O...O) = 2.53$ Å, is merely another mathematically possible solution. The importance of the utilization of the microwave spectroscopic information in this structure analysis is best demonstrated by the following statement: had we had a starting parameter set corresponding to the vicinity of minimum "f", the true parameter set would have never been obtained as was shown for analogous situations by Hamilton (1964; 1965).

Vinyl sulphonyl chloride, $CH_2{=}CH{-}SO_2Cl$. So far only electron diffraction has been employed (Brunvoll and Hargittai, 1977) to elucidate its molecular structure. We are aware of no vibrational spectroscopic study on this compound, and spectroscopic calculations based on a constructed force field from those of analogous molecules were used to produce mean amplitudes of vibration to aid the electron diffraction structure analysis. They are presented in Tables 13 and 14.

A number of conformers are a priori possible for vinyl sulphonyl chloride. They can be characterized by the forms shown in Figure 15. There are two or may be even three forms present according to the electron diffraction data. They are characterized by angles of rotation around the S-C bonds of 60, 80, and 160–180°

Table 13

Calculated mean amplitudes of vibration (Å) for bonds and some rotation-independent non-bond distances of vinyl sulphonyl chloride and benzene sulphonyl chloride

Bonds			Non-bonded atomic pairs			
	CH_2CHSO_2Cl (a)	$C_6H_5SO_2Cl$ (b)	CH_2CHSO_2Cl (a)		$C_6H_5SO_2Cl$ (b)	
S=O	0.036	0.036	S...Cl	0.061	S...C2	0.070
S-Cl	0.051	0.054	Cl...C2	0.064	Cl...Cl	0.070
S-C	0.044	0.045	O...C2	0.066	O...Cl	0.069
C-C	0.047	0.040	O...O	0.069	O...O	0.074
C-H	0.078	0.077	Cl...O	0.076	Cl...O	0.084
					Cl...C3	0.057
					Cl...C4	0.062
					S...C3	0.071
					S...C4	0.068

(a) At 333 K, the numbering of atoms is given in Figure 15, Brunvoll and Hargittai (1977)

(b) At 411 K, the numbering of atoms is given in Figure 21, Brunvoll and Hargittai (1976)

if $0°$ angle is assigned to the anti form. The most abundant form is with the rotation angle around $60°$. A detailed elucidation of the conformational properties is hindered by the relatively small

Table 14

Calculated mean amplitudes of vibration (Å) for distances
dependent on rotation around the S-C bond in vinyl sulphonyl
chloride[*]

	$\tau = 0^{\circ}$	60°	90°	180°
$Cl...C$	0.068	0.084	0.095	0.100
$O8...Cl$	0.093	0.103	0.104	0.086
$O9...Cl$	0.093	0.075	0.068	0.086
$Cl...H3$	0.134	0.172	0.198	0.176
$Cl...H4$	0.118	0.133	0.141	0.131
$Cl...H5$	0.129	0.145	0.141	0.097
$O8...H3$	0.184	0.180	0.185	0.169
$O8...H4$	0.128	0.127	0.129	0.126
$O8...H5$	0.123	0.098	0.100	0.131
$O9...H3$	0.184	0.155	0.140	0.169
$O9...H4$	0.128	0.118	0.113	0.126
$O9...H5$	0.123	0.137	0.134	0.131

[*] At 333 K, the numbering of atoms is given
in Figure 15, Brunvoll and Hargittai (1977)

contribution of the rotation-dependent interactions to the total
scattering. Figure 16 shows the experimental radial distribution.
The rotation-dependent distances are not expected to appear at
the regions below $r \sim 2.9$ Å. A further difficulty in the

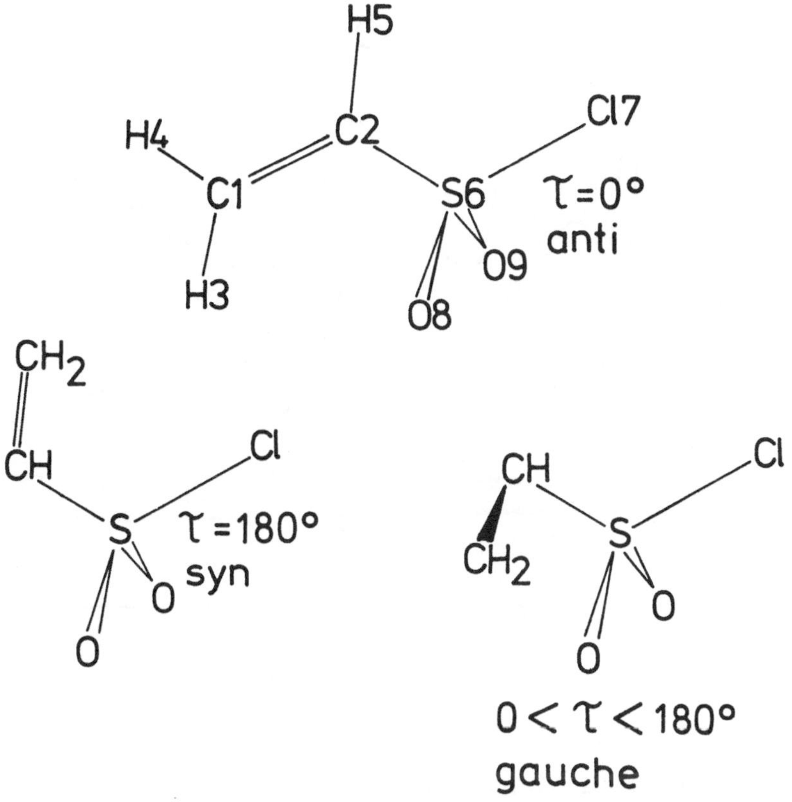

Figure 15

Molecular models of vinyl sulphonyl chloride

conformational analysis is that the contribution from rotation-
dependent distances is relatively insensitive to the changes in
the angle of rotation in its relatively large intervals. This
insensitivity comes not only from the insensitivity of the

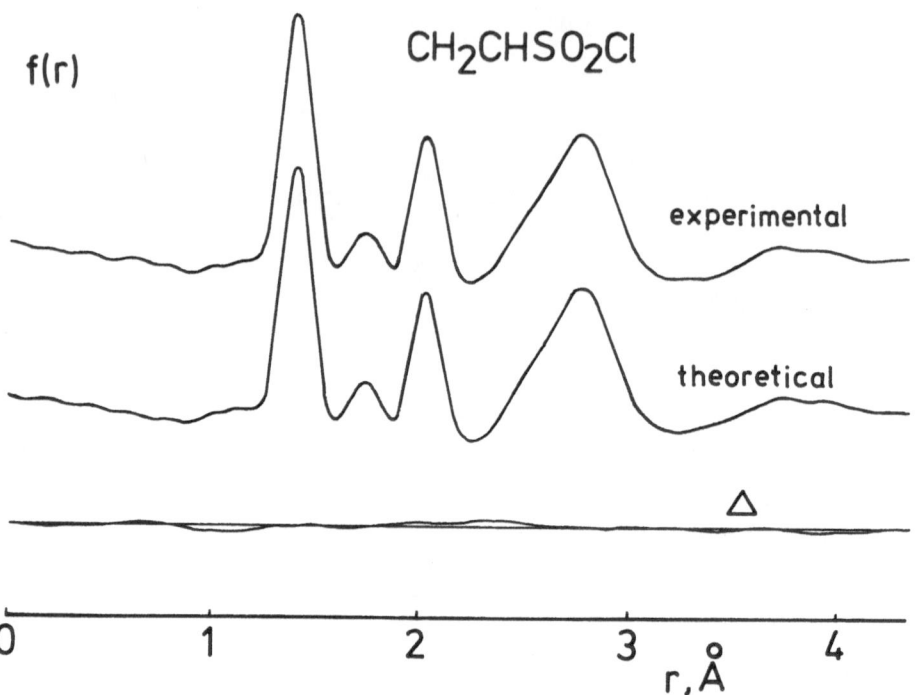

Figure 16

Radial distributions of vinyl sulphonyl chloride

The theoretical curve was calculated for a mixture of conformers:

amount (%)   angle of rotation ($^{\circ}$)

45   58

38   167

17   80

The geometrical and vibrational parameters are given in Tables

13-15

rotation-dependent distances to the rotation angle but also from the fact that the various rotation-dependent distances exchange their place of appearance on the radial distribution curve. This point is illustrated by Figure 17. On the other hand, it may be

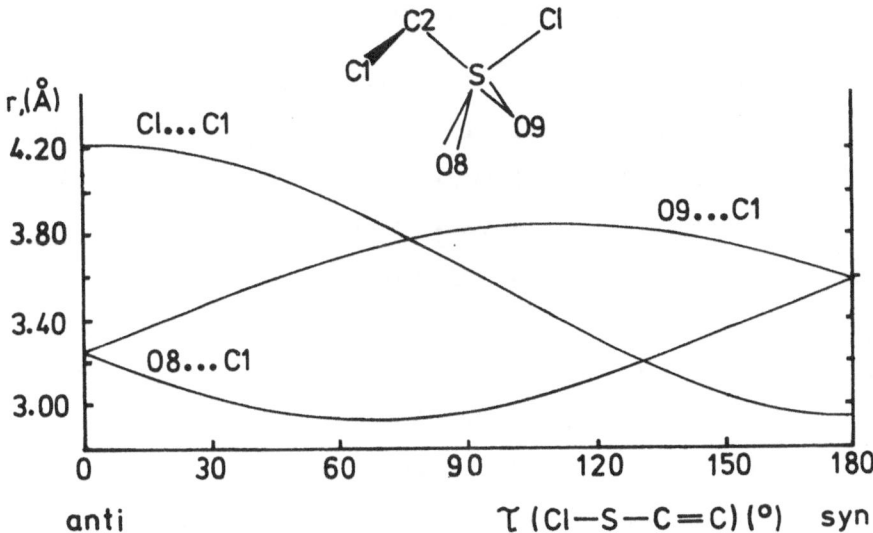

Figure 17

The variations of the most important rotation-dependent distances as a function of the angle of rotation in vinyl sulphonyl chloride

stressed that the geometrical parameters (bond distances and bond angles, given in Table 15) and also the bond amplitudes of vibration were very stable under various conditions used in the refinements and very insensitive to the conformational properties

Table 15

Bond lengths (Å) and bond angles (°) of vinyl sulphonyl chloride and benzene sulphonyl chloride from electron diffraction

	CH$_2$=CHSO$_2$Cl (a)	C$_6$H$_5$SO$_2$Cl (b)
r$_a$(S=O)	1.420±0.006	1.417±0.012
r$_a$(S-Cl)	2.035±0.005	2,047±0.008
r$_a$(S-C)	1.744±0.005	1.764±0.009
r$_a$(C-C)	1.357±0.018	1.403±0.010
∠ C-S-Cl	100.2±0.6	100.9±2.0
∠ C-S=O	109.8±0.4	110.0±2.5
∠ O=S=O	122.0±1.0	122.5±3.6
∠ Cl-S=O	106.4±0.5	105.5±1.8
∠ C-C-S	121.3±1.7	

(a) Brunvoll and Hargittai (1977)

(b) Brunvoll and Hargittai (1976)

of the models for which their refinements were carried out.

Divinyl sulphone, (CH$_2$=CH)$_2$SO$_2$. The electron diffraction structure analysis was augmented by a detailed experimental vibrational spectroscopic investigation and semiempirical CNDO/2 molecular orbital calculations (Hargittai, Rozsondai, Nagel,

Bulcke, Robinet and Labarre, 1977). Two distinct features in the
outer part of the experimental radial distribution (Figure 18)

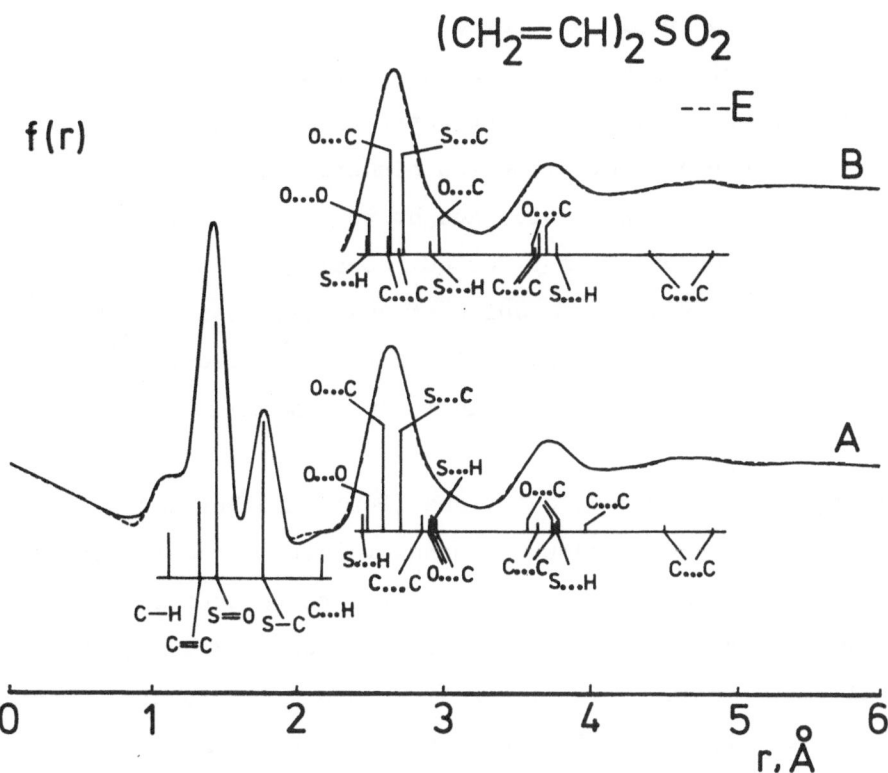

Figure 18

Radial distributions of divinyl sulphone. The solid lines were
computed for models A and B which differ in the conformational
properties as well as in geometrical and vibrational parameters
(see text)

could be attributed to contributions from interactions between
terminal carbon atoms and thus they were direct indications of two
rotational isomers. They could be forms with C_s ($C...C \sim 4.55$ Å)
and C_2 ($C...C \sim 4.85$ Å) symmetry. Considering, however, the two
axes of rotation in divinyl sulphone, there may be numerous
conformations of which only a few are shown in Figure 19. The
contributions from the different rotational forms are expected to
overlap on the radial distribution hindering their identification.
The CNDO/2 calculations[*] were thought to be useful mainly to
bring up more conformers for further consideration in the electron
diffraction analysis. The results of the calculations are illustrated by the isoenergy curves of Figure 20. Three forms proved
to have minimum energy. The forms with symmetries C_2, C_s, and C_1

---

[*] Semiempirical CNDO/2 calculations (Pople and Beveridge, 1970)
have been seen to produce erroneous results for delocalized
systems, whereas they have been proved to be useful for
numerous localized systems (see e.g. Gropen and Seip, 1971;
Perahia and Pullman, 1973; Corosine, Crasnier, Labarre, Labarre
and Leibovici, 1973). The bond lengths determined by electron
diffraction for divinyl sulphone did not indicate any
appreciable delocalization.

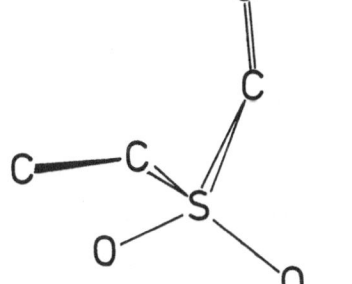

Figure 19

Some of the possible rotational forms of divinyl sulphone. The three forms are characterized by the following angles of rotation:

	$\varphi_1 (^\circ)$	$\varphi_2 (^\circ)$
C_2	60	60
C_s	60	300
C_1	60	180

have statistical weights 2, 2, and 4, respectively. The barriers between the minima are relatively large. The electron diffraction experimental data could be approximated fairly well by mixtures of the C_2 and C_s conformers. The agreement between the experi-

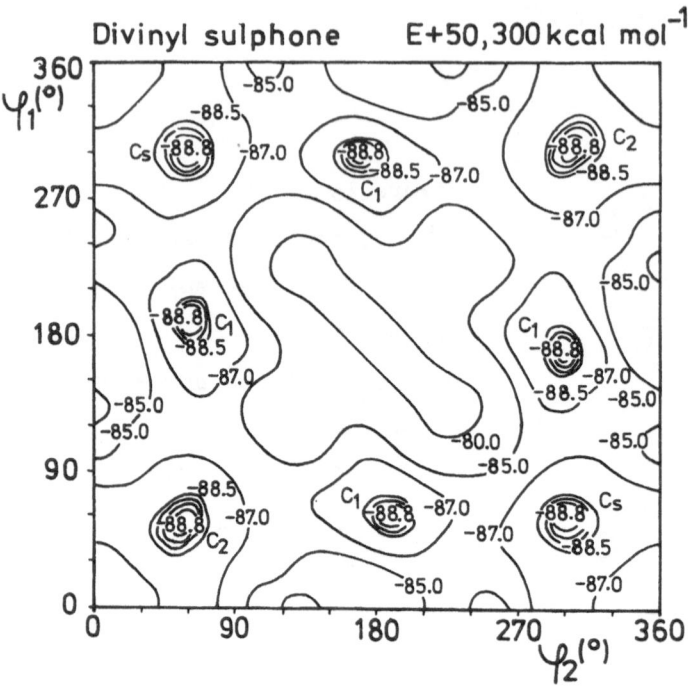

Figure 20

Isoenergy curves calculated by the CNDO/2 method for different
combinations of the angles of rotation in divinyl sulphone

mental and theoretical distributions could be further improved,
however, by adding some C_1 form to the mixture. Another approach
in improving the agreement was to remove the constraints on the
rotation angles which had been applied in order to preserve
symmetry. Unfortunately, some of the geometrical parameters were
sensitive to changes in the conformational properties of the

models used in the refinements. Further uncertainty was caused by the lack of knowledge concerning the mean amplitudes of vibration. Two structural parameter sets were used to characterize the results obtained in various refinement schemes. In one (A), only C_2 and C_1 conformers were present, but the C_1 form as was obtained in the refinement, could be considered as a distorted C_s form. This parameter set was also characterized by mean amplitudes of vibration for non-bond distances around the sulphur atom which were very similar to those calculated for analogous molecules (e.g. $CH_2=CH-SO_2Cl$). The other set (B) was obtained for a model containing three forms ($C_2 + C_s + C_1$) and was characterized by somewhat larger than expected amplitudes for the non-bond distances around the sulphur. The most striking difference between the parameter sets A and B occured in the C–S–C angle (107.7^o and 95.0^o, respectively), and on the basis of statistical criteria it is not possible to choose between them. They are listed in Table 16 together with values obtained for $\angle$ C–S–C and $\angle$ O=S–C referring to models A and B. The latter two parameters cannot be considered established from this analysis. The O...O distance was assumed in most of the refinements but the effect of this assumption has been considered (as in case of CH_3OSO_2Cl, for example) in the error estimation. Of the geometrical parameters of divinyl sulphone, what was really well determined in this study were the following parameters: $r(S=O)$, $r(S-C)$, $r(C=C)$, $r(C-H)$, and $\angle$ S–C=C.

Table 16

Bond lengths (Å) and bond angles (°) of divinyl sulphone from
electron diffraction[*]

$(CH_2=CH)_2SO_2$			
r_a(S=O)	1.438±0.003	∠ O=S=O	119.5±1.2
r_a(S-C)	1.769±0.004	∠ O=S-C	107 (A)[**] – 110 (B)[**]
r_a(C=C)	1.332±0.004	∠ C-S-C	108 (A)[**] – 95 (B)[**]
		∠ S-C=C	121.5±0.3

[*] Hargittai, Rozsondai, Nagel, Bulcke, Robinet and Labarre
   (1977)
[**] See in the text

Benzene sulphonyl chloride, $C_6H_5SO_2Cl$. The electron
diffraction structure analysis utilized vibrational spectroscopic
data as much as possible (Brunvoll and Hargittai, 1975). Since
this molecule was the first sulphone for which this was done in
a more complete way, the procedure is described here in somewhat
more detail.

The purpose of the spectroscopic calculations was to provide
mean parallel, l and mean square perpendicular, $\langle(\Delta x)^2\rangle$ and
$\langle(\Delta y)^2\rangle$ amplitudes of vibration or rather, the K values. The
spectroscopic l values were used in the least-squares refinement
and for comparison with the results obtained from the electron
diffraction data. The l values used in the refinement were

needed as initial values only, or as assumed parameters
throughout the analysis for atomic pairs with relatively small
contribution to the total scattering. The spectroscopic
calculations were based on the force fields of benzene (Brooks
and Cyvin, 1962; Brooks, Cyvin and Kvande, 1965) and of the
SO_2Cl fragment of methane sulphonyl chloride (Cyvin, Dobos,
Hargittai, Hargittai and Augdahl, 1973), and also on the experimental frequencies of benzene sulphonyl chloride (Ham, Hambly
and Laby, 1960; Uno, Machida and Hanai, 1968). The force field
of benzene sulphonyl chloride produced this way can be considered
tentative only since the molecule is of low symmetry and many
frequencies are lacking. Nevertheless, the calculated values
proved to be very useful in the structure analysis. The mean
(parallel) amplitudes of vibration for the most important atomic
pairs are compiled in Table 13 (rotation-independent distances)
and in Table 17 (rotation-dependent distances). The numbering of
atoms is presented in Figure 21. Calculated K values were used
in converting the r_a distances into r_α distances. The conversion
was needed in the conformational analysis, primarily in trying to
facilitate a decision between a symmetrical and an asymmetrical
model (described in more detail below) so as to minimize the
effect of intramolecular motion on the results of the structure
analysis. Figure 22 shows how the vibrational corrections
were applied in the course of the refinements.

The rotational forms of benzene sulphonyl chloride
considered in this analysis are presented in Figure 23. One of
the two symmetrical forms, viz. the one in which the plane of the

Table 17

Calculated mean amplitudes of vibration (Å) for distances
dependent on rotation around the S-C bond in benzene
sulphonyl chloride[*] (interactions involving hydrogen are
not listed)

	$\tau = 0°$	40°	76.3°	90°
C(6)...Cl	0.105	0.128	0.134	0.130
C(6)...O(15)	0.107	0.081	0.073	0.076
C(6)...O(14)	0.107	0.121	0.124	0.123
C(5)...Cl	0.116	0.139	0.142	0.138
C(5)...O(15)	0.119	0.091	0.083	0.085
C(5)...O(14)	0.119	0.126	0.128	0.128
C(4)...Cl	0.102	0.125	0.135	0.136
C(4)...O(15)	0.127	0.109	0.106	0.108
C(4)...O(14)	0.127	0.119	0.110	0.108
C(3)...Cl	0.086	0.107	0.132	0.138
C(3)...O(15)	0.127	0.124	0.127	0.128
C(3)...O(14)	0.127	0.111	0.091	0.085
C(2)...Cl	0.080	0.096	0.123	0.130
C(2)...O(15)	0.116	0.118	0.121	0.123
C(2)...O(14)	0.116	0.101	0.081	0.076

[*] At 411 K, the numbering of atoms is given in Figure 21,
Brunvoll and Hargittai (1976)

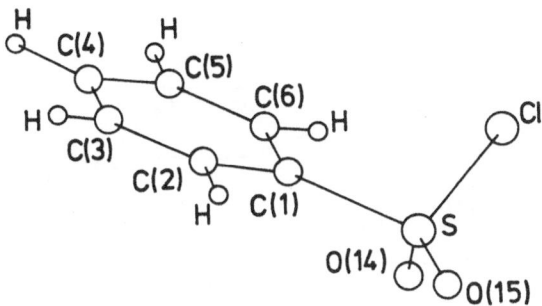

Figure 21

The molecular model of benzene sulphonyl chloride and the
numbering of atoms of the C_6SO_2Cl skeleton

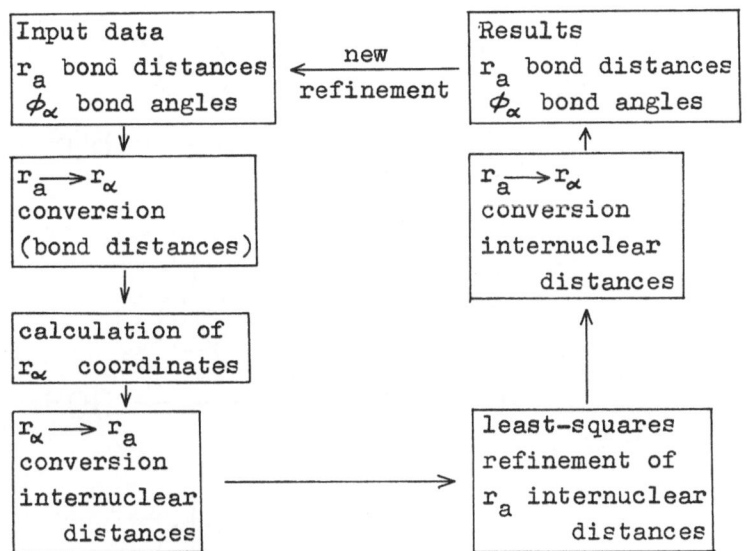

Figure 22

The application of vibrational corrections in the electron
diffraction structure analysis

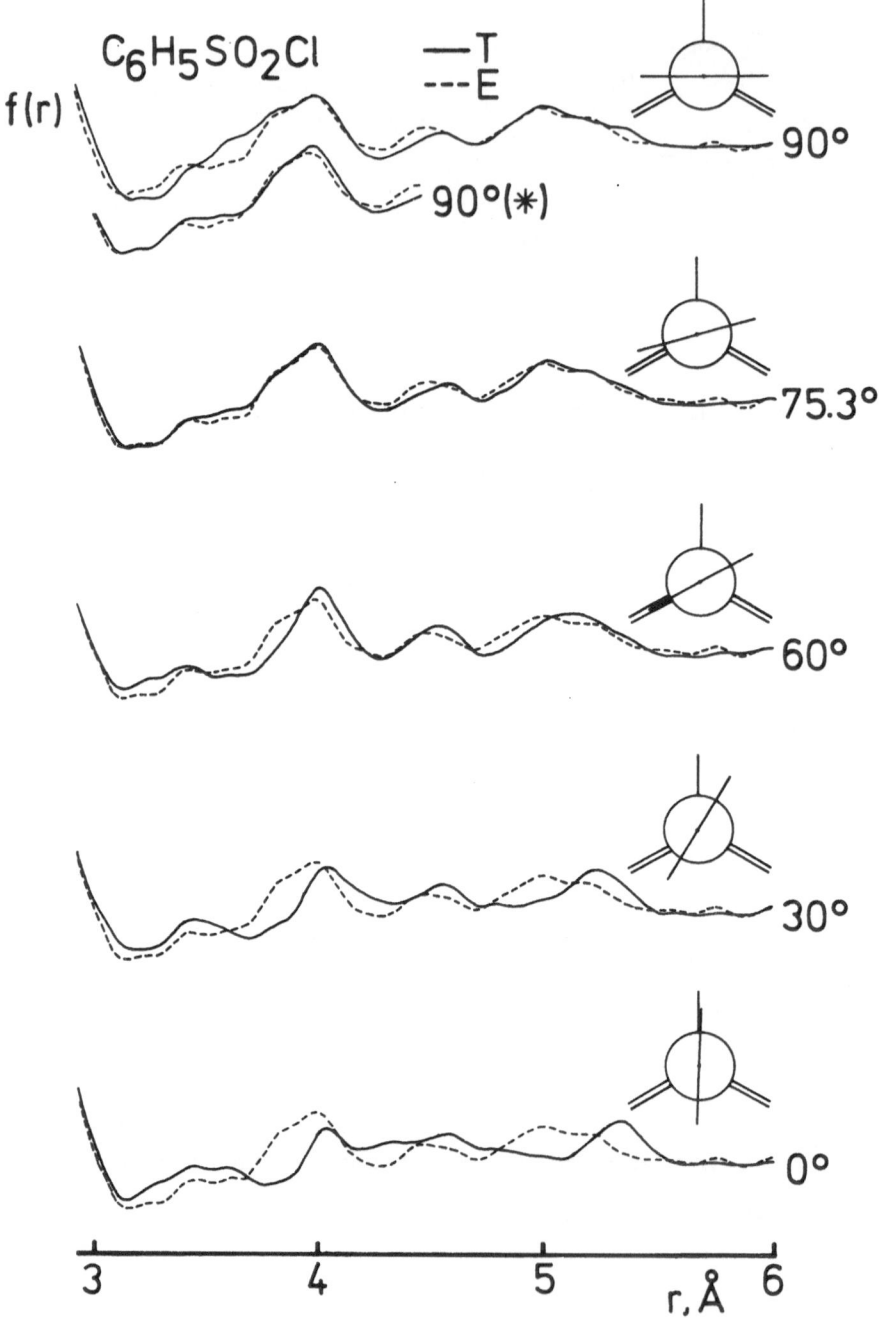

Figure 23

Newman projections of rotational forms of benzene sulphonyl chloride representing view along the C-S bond and the angles of rotation. The rotation-dependent portions of the radial distributions are shown, E - experimental, and T - theoretical. The parameters used for calculating the theoretical curves are presented in Tables 13, 15 and 17, except for the curve 90°(*) for which some larger mean amplitudes were used as obtained in the electron diffraction analysis (most notably $\ell(C2...C\ell)$ = 0.258 Å)

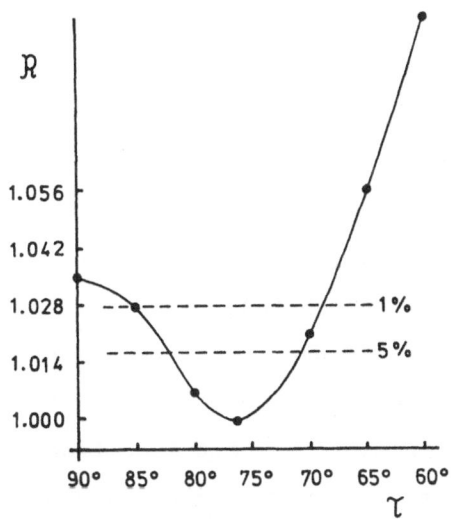

Figure 24

The R factor ratio versus rotational angle τ showing 5% and 1% significance levels for deviation from $\tau = 90^\circ$ (cf. Hamilton, 1965) in benzene sulphonyl chloride

benzene ring bisects the O=S=O angle and eclipses the sulphur-
chlorine bond ($\tau = 0^{o}$), could be excluded. The experimental data
could be approximated equally well by a model in which the plane
of the benzene ring is perpendicular to the plane containing the
sulphur-chlorine bond and the line bisecting the O=S=O angle
($\tau = 90^{o}$) or an asymmetric model with $\tau = 75^{o}$. In the former
model particularly large vibrational amplitude was needed for the
shortest rotation-dependent carbon...chlorine distance, while in
the latter the spectroscopically calculated mean amplitudes of
vibration were utilized. The rotation-dependent parts of the
radial distribution curves for various angles of rotation are
reproduced in Figure 23.

It was attempted to choose between the two models on the
basis of r_{α} structures. However, the asymmetric model showed no
appreciable change. The angle of rotation was obtained again 75°,
and the deviation from the symmetric structure was present even at
a 99% significance level according to a Hamilton test (Hamilton,
1964; 1965), as seen in Figure 24. A critical remark has to be
made here, however, concerning the procedure for vibrational
corrections. The correction terms were calculated, as usual, in
the harmonic approximation. The $r_a \rightarrow r_{\alpha}$ conversion is especially
important in case of large amplitude motion where anharmonicity
can no longer be ignored. The r_{α} parameters are not free from
the effect of anharmonicity, $r_{\alpha} \approx r_e + \langle \Delta z \rangle$, and the above
procedure is supporting the notion about the asymmetric model to
the extent that only small vibrations take place.

The bond lengths and bond angles determined in benzene

sulphonyl chloride are collected in Table 15.

N,N-dimethyl sulphamoyl chloride, $(CH_3)_2NSO_2Cl$. A recent
electron diffraction reinvestigation (Hargittai and Brunvoll,
1976) confirmed the conformation suggested by an earlier study
(Vilkov and Hargittai, 1967). The molecule has C_s symmetry and
the N-C bonds stagger the sulphur-chlorine bond. This form (I)
and two others (II and III) which were excluded are shown by their
Newman projection in Figure 25 with the corresponding radial
distributions. Slight deviations from the C_s symmetry in form I
(characterized by an angle of rotation around the S-N bond up to
about 4°) somewhat further improved the agreement. This was seen
not so much in the changes of the generalized R values but in
the regions of the radial distributions where the rotation-
dependent distances occur. The other geometrical parameters,
given in Table 18, showed no appreciable changes in these calculations.

Spectroscopic calculations provided mean amplitudes of
vibration for the electron diffraction structure analysis. These
calculations were based on a force field constructed from those
of trimethyl amine (Gebhardt, 1971) and the SO_2Cl fragment of
methane sulphonyl chloride (Cyvin, Dobos, Hargittai, Hargittai
and Augdahl, 1973). This force field was then modified in order
to reproduce the experimentally determined vibrational frequencies (Bürger, Burczyk, Blascheffe and Safari, 1971). Some of the
mean amplitudes of vibration as calculated or determined from the
electron diffraction data are given in Table 18.

Table 18

Bond lengths ($\overset{o}{A}$), bond angles (^{o}) and mean amplitudes of vibration ($\overset{o}{A}$) in N,N-dimethyl sulphamoyl chloride from electron diffraction (Hargittai and Brunvoll, 1976)

	r	ℓ
S=O	1.421±0.004 $\overset{o}{A}$	0.038±0.004 $\overset{o}{A}$
S-Cl	2.064±0.005 $\overset{o}{A}$	0.061±0.003 $\overset{o}{A}$
S-N	1.618±0.005 $\overset{o}{A}$	0.051±0.006 $\overset{o}{A}$
C-N	1.481±0.012 $\overset{o}{A}$	0.050* $\overset{o}{A}$
C-H	1.096±0.010 $\overset{o}{A}$	0.079±0.010 $\overset{o}{A}$
N-S-Cl	103.0±0.5^{o}	
N-S=O	108.8±1.4^{o}	
S-N-C	115.8±0.7^{o}	
C-N-C	114.6±2.2^{o}	
O=S=O	122.7±2.3^{o}	
O=S-Cl	105.8±0.4^{o}	
C...Cl		0.17±0.06 $\overset{o}{A}$
(O...C) short		0.155* $\overset{o}{A}$
(O...C) long		0.08±0.04 $\overset{o}{A}$

* From spectroscopic calculations

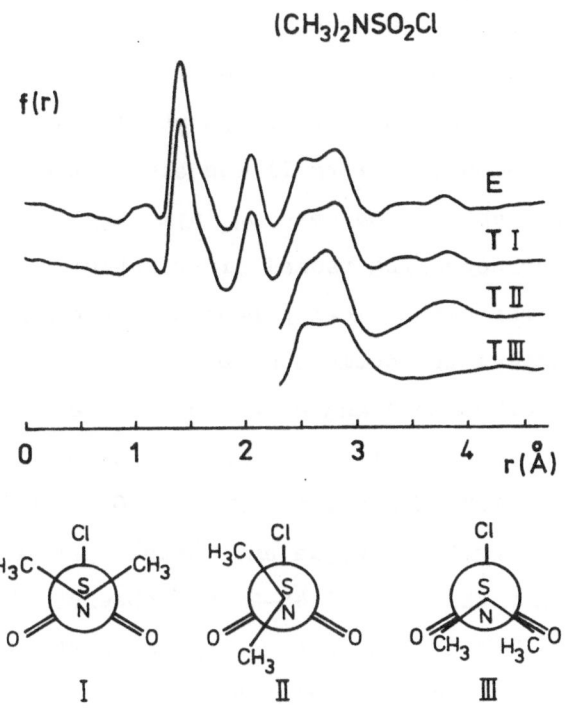

Figure 25

Radial distributions of N,N-dimethyl sulphamoyl chloride.
E - experimental, T - calculated for the corresponding
conformers presented by Newman projections representing
view along the S-N bond

$\underline{\text{N,N'-tetramethyl sulphonyl diamine}}$, $(CH_3)_2NSO_2N(CH_3)_2$.
Two rotational isomers were found to coexist in the vapour
(Hargittai, Vajda and Szőke, 1973). They are shown by their

Newman projection in Figure 26. The prevailing form in the vapour phase is the same as the one found in the crystal phase by X-ray diffraction (Jordan, Smith, Lohr and Lispcomb, 1963). In addition to form I, a smaller amount of form II was also detected by electron diffraction. Both forms have C_{2v} symmetry and their ratio is 4:1. The corresponding radial distributions are also shown in Figure 26. The geometrical parameters determined by the two cited investigations are collected in Table 19.

Concerning the O=S=O bond angle, or rather the r(O...O) parameter from the electron diffraction study, a comment has to be made. The value of 2.42 Å was given for the O...O distance in the original paper (Hargittai, Vajda and Szőke, 1973) which is considerably less than the usual 2.48–2.49 Å observed in a large series of molecules. The stated uncertainty was particularly large (σ = 0.038 Å) and the strong correlation between the parameters did not indeed allow a determination of this parameter. In subsequent calculations (Hargittai and Vajda, 1975) it was shown that assuming r(O...O) to be 2.48 Å, the experimental data could be reproduced equally well as previously while all the other parameters showed no appreciable changes.

Sulphonyl chloride isocyanate, $OCNSO_2Cl$. The electron diffraction data (Brunvoll, Hargittai and Seip, 1977) were found to be consistent with two models differing mainly in the relationship of the C=O and N=C bond lengths. The bond lengths and bond angles of the two models are presented in Table 20 together with the calculated mean amplitudes of vibration. As is

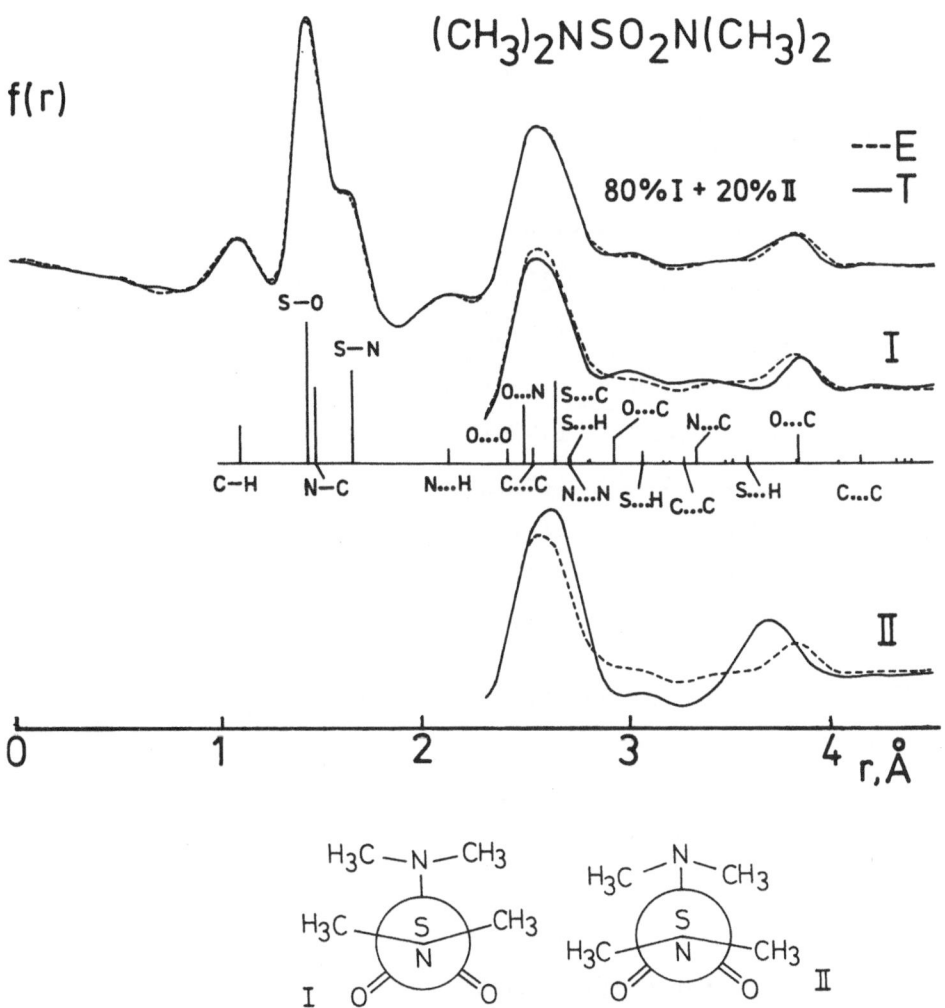

$(CH_3)_2NSO_2N(CH_3)_2$

f(r)

---E
—T

80% I + 20% II

I

II

Figure 26

Newman projections representing view along the S—N bond of
N,N'-tetramethyl sulphonyl diamine for the two conformers
found to coexist in the vapour with a 4:1 relative abundance.
The radial distributions (T) are shown for the individual
conformers and their mixture and also the experimental one (E)

Table 19

Bond lengths (Å) and bond angles ($^{\circ}$) in N,N'-tetramethyl
sulphonyl diamine in the vapour phase and in the crystal
phase

	(a)	(b)
r(S=O)	1.432±0.010	1.445±0.005
r(S-N)	1.651±0.003	1.623±0.005
r(C-N)	1.475±0.013	1.476±0.007
∠ S-N-C	115.2±1.1	118.8±0.4
∠ C-N-C	118.0±3.2	112.9±0.4
∠ N-S-N	110.5±1.3	112.6±0.4
∠ O=S=O	(*)	119.7±0.4

(a) Vapour phase, electron diffraction; Hargittai,
    Vajda and Szőke (1973); (*) see text
(b) Crystal phase, X-ray diffraction; Jordan, Smith,
    Lohr and Lipscomb (1963)

discussed in a later section on the "Geometrical variations in
the rest of the molecule", the N=C bond was observed to be longer
than the C=O bond in the majority of the isocyanate molecules
studied to date. Mainly on this basis, the parameter set (i) is
preferred. Note, however, that the lengths and vibrational
amplitudes of the bonds other than N=C and C=O are invariant to
the choice of the relationship between these two bond lengths.

Table 20

Molecular parameters of sulphonyl chloride isocyanate

(Brunvoll, Hargittai and Seip, 1977)

| | Geometrical parameters[a] | | Vibrational amplitudes[b] |
	Model(i)	Model(ii)	
Bonds (Å)			
C=O	1.159(19)	1.218	0.038
N=C	1.221(23)	1.152	0.036
S–N	1.656(3)	1.657	0.050
S–Cl	2.019(3)	2.018	0.052
S=O	1.417(3)	1.416	0.034
Angles (°)			
S–N–C	122.4(12)	125.8	
N–S=O	108.5(22)	108.7	
N–S–Cl	99(3)	98.2	
O=S–Cl	107.8	107.9	
O=S=O	122.4	122.7	
τ_1(Cl–S–N=C)	109(4)	103	
τ_2(Cl–S–N=C)	70(10)	63	
Amount of form with τ_1	69(7)%	67%	

The conformational ratio and most of the angles differ in the two models only within their estimated errors. The most notable change appears in the S–N=C bond angle.

The conformers characterized by rotation angles $109\pm4°$ and $70\pm10°$ were found to coexist in the vapour phase with a relative abundance of 2:1. The rotation angle $0°$ corresponds to a form in

Table 20 (continued)

Vibrational amplitudes $(\overset{\circ}{A})^b$ for non-bond distances

Rotation-independent   Rotation-dependent

				1	2
O1...N	0.043	O1...Cl		0.220	0.182
O1...S	0.094	O1...O6		0.087	0.115
C...S	0.070	O1...O7		0.192	0.197
N...Cl	0.088	C...Cl		0.154	0.132
N...O6	0.071	C...O6		0.069	0.089
Cl...O6	0.077	C...O7		0.130	0.133

---

[a] Distances r_a parameters, paranthesized values
estimated total errors;

[b] From spectroscopic calculations

which the S-Cl bond is anti to the O=C=N chain. This form could
be excluded on the basis of the radial distribution curves. A
least-squares test revealed the presence of no appreciable amount
of the form with rotation angle 180°. The Newman-projections of
the four rotamers mentioned are shown in Figure 27 together with
the calculated, for each of them, radial distributions and the
experimental curve. The distribution calculated for the 2:1
mixture of the two above mentioned forms is also shown. Note that
the two forms found to be present both possess planar Cl-S-N=C=O
chains.

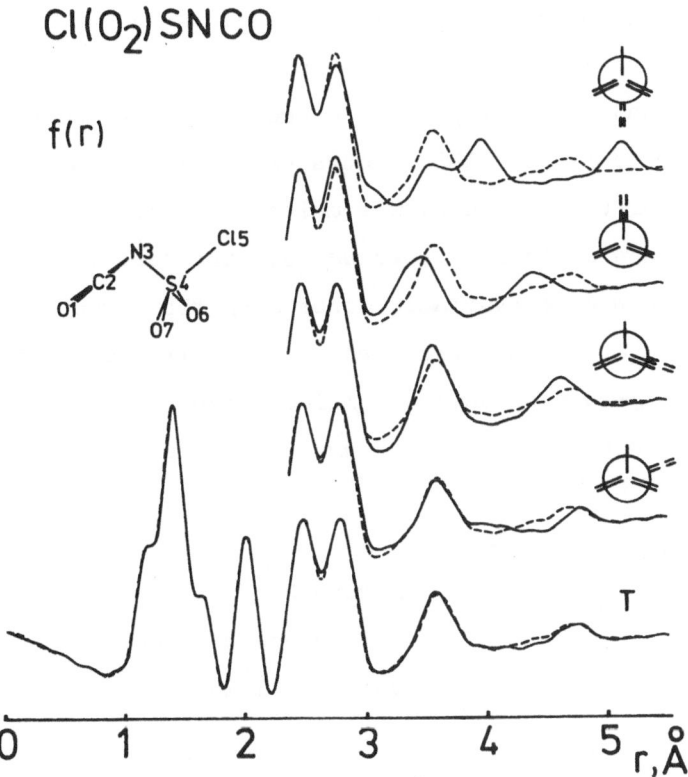

Figure 27

Newman projections of four rotamers of sulphonyl chloride
isocyanate representing view along the S-N bond. The two forms
with 109° and 70° rotation angles (0° corresponds to the form
in which the S-Cl bond is anti to the O=C=N chain) were found
to coexist in the vapour with a 2:1 relative abundance. The
radial distribution for this mixture (T) as well as for the
individual conformers are shown, together with the experimental
curves

Thiirane 1,1-dioxide (ethylene episulphone), $CH_2-CH_2-SO_2$.
The microwave spectra of three isotopic species (normal,
^{13}C-isotopic and ^{34}S-isotopic) have been investigated by Nakano,
Saito and Morino (1970). The spectra provided evidence for the
C_{2v} symmetry of the molecule. The molecular model is shown in
Figure 28. Since deuterated species have not been investigated,
no information became available concerning the structure of the
methylene groups. In the structure analysis it was assumed that
the methylene groups of thiirane 1,1-dioxide and thiirane
(ethylene episulphide) have the same structure. The following

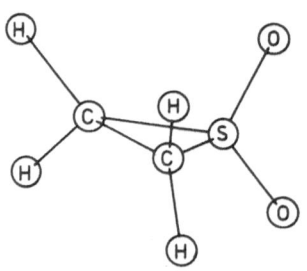

Figure 28

The molecular model of

thiirane 1,1-dioxide

parameters were determined in the latter (Saito, 1970):
r(C-H) = 1.078 Å, ∠H-C-H = 116°0', and ∠H-C-C = 151°43'.
The microwave spectrum of the ^{34}S-isotopic species added nothing
to the structural information since the sulphur atom is situated
near the centre of gravity of the molecule. The following four
parameters (r_0) were determined for the C_{2v} symmetry heavy atom
skeleton:

$$r(S=O) = 1.439 \pm 0.006 \text{ Å}$$
$$r(S-C) = 1.731 \pm 0.006 \text{ Å}$$
$$\angle O=S=O = 121°26' \pm 30'$$
$$\angle C-S-C = 54°40' \pm 15' \; .$$

On the basis of the above independent parameters, the C-C bond length was calculated to be 1.590 ± 0.011 Å !

<u>Thiete 1,1-dioxide (thiete sulphone)</u>, $CH{=}CH{-}CH_2{-}SO_2$.
In order to determine the O...O distance from the rotational constants, Ralowski, Ljunggren and Mjöberg (1973) assumed a planar ring configuration in their microwave spectroscopic investigation. This assumption was justified by the results of an X-ray diffraction study of this compound in the crystal phase (Lowenstein, 1965). The following bond lengths and bond angles were determined in the X-ray diffraction work:

$r(C{-}C)$	$= 1.52 \pm 0.04$ Å	$\angle C{-}S{-}C = 80.5°$
$r(C{=}C)$	$= 1.39 \pm 0.03$ Å	$\angle S{-}C{-}C = 85.1°$
$r({-}C{-}S)$	$= 1.79 \pm 0.04$ Å	$\angle S{-}C{=}C = 90.0°$
$r({=}C{-}S)$	$= 1.77 \pm 0.03$ Å	$\angle C{-}C{=}C = 104.5°$
$r(S{=}O)$	$= 1.43 \pm 0.02$ Å	$\angle O{=}S{=}O = 115.5°$

The value of 2.478 Å determined unambiguously for $r(O...O)$ from the microwave spectrum is not consistent with the pair of values determined in the X-ray diffraction study for $r(S{=}O)$ and $\angle O{=}S{=}O$. Accordingly, either $r(S{=}O)$ or $\angle O{=}S{=}O$ is too small. It seems to us that the latter is more probable and from $r(S{=}O) = 1.43$ Å and $r(O...O) = 2.48$ Å, $\angle O{=}S{=}O$ is estimated to be 120.1°.

<u>Tetrahydrothiophene 1,1-dioxide (tetramethylene sulphone)</u>, $CH_2{-}CH_2{-}CH_2{-}CH_2{-}SO_2$. An electron diffraction investigation (Naumov, Semashko and Shaidulin, 1973) showed an "envelope" and

a "half-chair" model (shown in Figure 29) to approximate equally

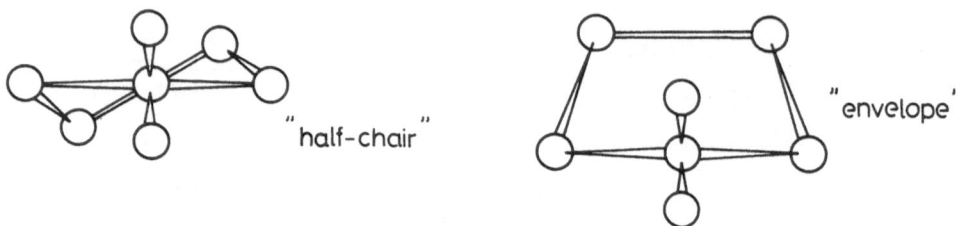

"half-chair"

"envelope"

Figure 29

The "half-chair" and "envelope" models of tetrahydrothiophene 1,1-dioxide

well the experimental data. A model with planar ring skeleton could be rejected. The conformational properties, however, could not be established unambiguously. On the other hand, the bond lengths and bond angles determined were not sensitive to the conformational choice of the model for which they were refined. The results obtained for the "half-chair" model are listed below:

$$r(CC-CC) = 1.533\pm0.015 \text{ Å}$$
$$r(SC-CC) = 1.540\pm0.015 \text{ Å}$$
$$r(S=O) \quad = 1.449\pm0.010 \text{ Å}$$
$$r(S-C) \quad = 1.798\pm0.008 \text{ Å}$$
$$\angle O=S=O = 115\pm3°$$
$$\angle C-S-C = 101.1\pm1.5°$$
$$\angle C-C-S = 104.3\pm1.0°$$
$$\angle C-C-C = 112.1\pm1.5°$$

The cited paper communicated mean amplitudes of vibration for the bonds only.

Table 21

Vapour-phase data on the SO_2 and SO_3 molecules

	SO_2		SO_3	
	(a)	(b)	(a)	(b)
r_g(S=O), Å	1.432		1.419	
r_o(S=O), Å				1.4198[*]
r_e(S=O), Å	1.427	1.4308	1.414	1.4184[**]
r_e(O...O), Å	2.453	2.4700	2.450	
$\angle_e$O=S=O	118.5°	119.3°		

(a) Electron diffraction; Clark and Beagley (1971)

(b) Rotational spectroscopy; SO_2 Morino, Kikuchi, Saito and Hirota (1964), [*] Kaldor and Maki (1973); [**] Dorney, Hoy and Mills (1973)

In conclusion of our surveying the experimental determinations of sulphone molecular geometries, the geometrical parameters of the SO_2 and SO_3 molecules are compiled in Table 21.

Part Two: STRUCTURAL VARIATIONS

Conformational properties

Figure 30 presents a summary of our findings concerning the
prevailing or characteristic rotational isomers in the vapour
phase of sulphone molecules. Looking for a characteristic pattern,
the following observation may be made. Staggered forms prevail
while eclipsed forms appear to be predominant when two double
bonds (one of them S=O) may take a syn orientation. This is a very
rough qualitative statement only, however, and there remain many
important questions unanswered. It may be of interest to list
some of them.

Why is the conformational behaviour of fluorosulphuric acid
methyl ester and chlorosulphuric acid methyl ester different?
The importance of van der Waals interactions has been examined
(Schultz, 1976) and no characteristic difference was revealed.

Are the symmetric and asymmetric forms observed for benzene
sulphonyl chloride different isomers indeed and if so which of
them, if only one, is the true conformer? It may be argued that
the stabilization of the asymmetric structure is facilitated by
intramolecular H...O interaction (cf. Figure 21). However, such
an interaction obviously cannot be present in the molecule of
benzyl chloride, $C_6H_5CH_2Cl$, where similar asymmetry was observed

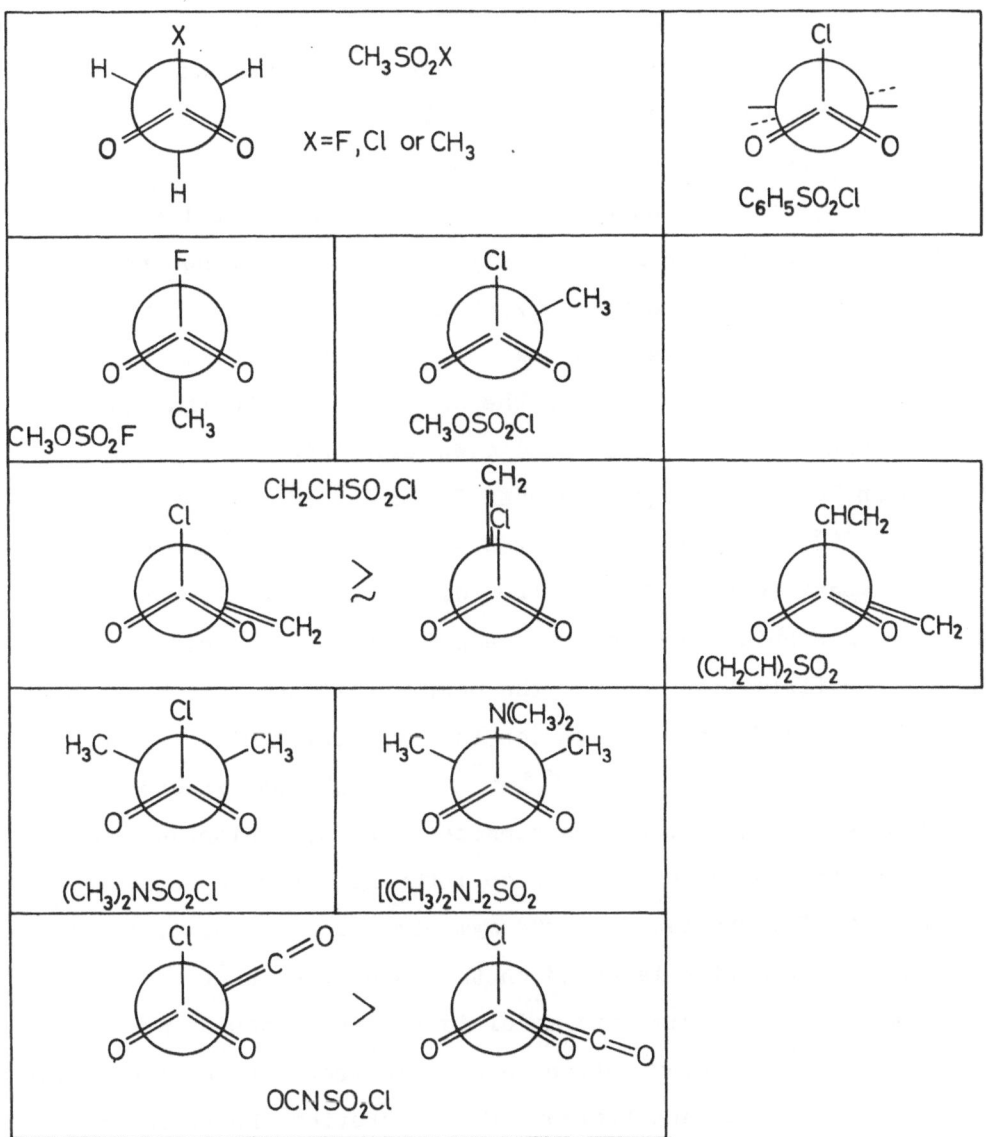

Figure 30

Summary of the conformational properties of sulphone molecules
in the vapour phase

by electron diffraction (Sadova, Vilkov, Hargittai and Brunvoll, 1976). It is also interesting that even more pronounced asymmetries have been observed in analogous systems, e.g. in $C_6H_5SO_2NSO_2C_6H_5$ (Cotton and Stokely, 1970) in the crystalline phase. Here, however, packing forces may strongly influence the molecular geometry. We have estimated the root mean square amplitude of the torsional angle, $\langle \Delta \tau^2 \rangle^{1/2}$, around the S–C bond from spectroscopic data (Brunvoll and Hargittai, 1976). Using the symmetric model ($\tau = 90^\circ$) and the same force field that was used in obtaining the (spectroscopic) mean amplitudes of vibration listed in Table 13, values between 8.1 and 8.5° were obtained for $\langle \Delta \tau^2 \rangle^{1/2}$. They seem to be too small for having the symmetric model compatible with the electron diffraction results provided the spectroscopically calculated amplitudes are assumed. However, if a force field, reproducing the mean amplitudes determined from the electron diffraction data for the symmetric model, would be used, a considerably larger $\langle \Delta \tau^2 \rangle^{1/2}$ could be expected. A final note in connection with benzene sulphonyl chloride concerns the estimated height of barrier to internal rotation obtained from the amount of deviation from the symmetric conformation. A method of such an estimation is based on the assumption that the appearance of the asymmetric model is a consequence of averaging over the intramolecular motion in the electron diffraction elucidation of the structure (Vilkov, Penionzhkevich, Brunvoll and Hargittai, 1977). A value of about 4 kcal mol^{-1} was obtained which does not seem to be unrealistic. Further physical evidence will be needed, however, to settle the question about the conformational choice of benzene sulphonyl chloride.

The prevailing form of internal rotation of N,N'-tetramethyl sulphonyl diamine is the same in the vapour and in the crystalline phases. Another form was also detected, however, in the vapour phase which can be obtained by 180° rotation around the S-N bond or by inversion of the nitrogen bonds, from the prevailing form. The molecules of thiobis(dimethylamine), $(CH_3)_2NSN(CH_3)_2$, were found to display similar conformational properties in the vapour phase (Hargittai and Hargittai, 1973b). As for the rotational isomerism of N,N'-tetramethyl sulphinyl diamine, $(CH_3)_2NSON(CH_3)_2$, three probable forms have been suggested (Hargittai and Vilkov, 1970), and the most important of them seemed to be analogous to the forms mentioned above. They are all depicted in Figure 31

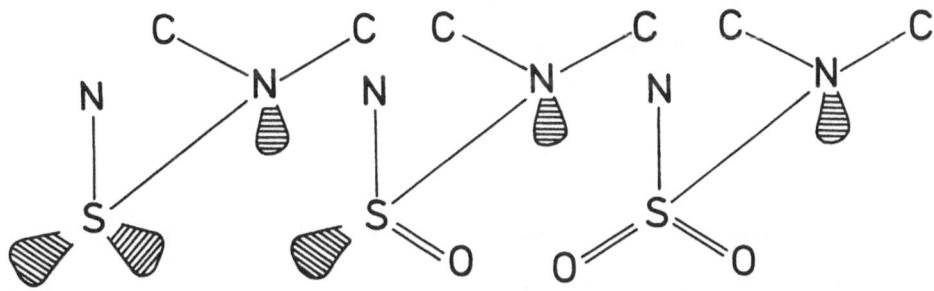

Figure 31

Ethane-like forms of the $NSNC_2$, $NSONC_2$, and NSO_2NC_2 skeletons as regards rotation around the S-N bond in $(CH_3)_2NSN(CH_3)_2$, $(CH_3)_2NSON(CH_3)_2$, and $(CH_3)_2NSO_2N(CH_3)_2$

showing also the lone pairs of electrons at nitrogen and sulphur suggesting that the prevailing forms are analogous to the staggered ethane-like structures. The only form determined for N,N-dimethyl sulphamoyl chloride is completely analogous to the prevailing forms of the other S-N derivatives (see Figures 30, 31).

Both forms found to be present in the vapour of sulphonyl chloride isocyanate have a planar (within experimental error) five-member chain

The coplanarity of some simple isocyanate molecules is an automatic consequence of the linearity of the O=C=N group, for example, OCNCN (Hocking and Gerry, 1976)

The benzene ring and the N=C=O chain are coplanar in phenyl isocyanate (Bouchy and Roussy, 1973)

Other isocyanate, and, for that matter, isothiocyanate
derivatives show a great variety of conformational properties as
was surveyed recently (Hargittai and Paul, 1977). Here only two
examples are given. Neither the staggered nor the eclipsed conformations (both with C_s symmetry) for trichlorosilyl isocyanate gave
satisfactory agreement with the electron diffraction experimental
data (Hilderbrandt and Bauer, 1969). A rotation angle of $24 \pm 4^{\circ}$
relative to the eclipsed form and a tilt angle of $5 \pm 1^{\circ}$ between
the threefold symmetry axis of the trichlorosilyl group and the
Si-N bond were determined for a model giving the best agreement
with the experiment. There were three comformations approximating
well the experimental electron diffraction data on $Cl_2(O)PNCO$
with P-N=C bond angles around $120-125^{\circ}$ (Naumov, Semashko and
Shatrukov, 1973). In none of the three forms is the N=C=O group
eclipsing a P-Cl or P=O bond.

Characteristic variations in the bond angles and bond lengths

In this section the geometrical parameters determined for the
sulphur bond configurations in the sulphone molecules are compared,
characteristic tendencies in their changes are pointed out, and
explanations are sought employing considerations on valence shell
electron pair repulsions and on non-bonded (atom-atom) interactions.
The role of electron pair repulsions is examined for the changes

in various series of parameters referring to molecules which differ from each other in the ligands (and their electronegativity) attached to the SO_2 group. The variations of non-bond distances are inspected in order to judge the possible importance of non-bonded interactions.

The valence shell electron pair repulsion (VSEPR) model (Gillespie, 1972; Gillespie and Nyholm, 1957; Sidgwick and Powell, 1940) is successfully used for explaining and accounting for variations in the molecular geometries of extensive classes of inorganic compounds. According to the VSEPR model, the bond configuration of an atom in a molecule is determined by the number of electron pairs in its valence shell, their relative size and shape. The relative orientation of the electron pairs in the valence shell is such as to maximize the distances between them. The VSEPR model summarizes the effects influencing the bond configuration in the following four rules: (i) A lone pair of electrons has larger space requirement and exercises stronger repulsion towards neighbouring electron pairs than a bonding pair. (ii) As the ligand electronegativity increases, more electron density is drawn away from the valence shell of the central atom and the repulsion from the bonding pair decreases. This rule will be referred to as "the electronegativity rule" in subsequent discussion. (iii) Multiple bonds have larger space requirement and exercise stronger repulsion than single bonds. This rule will be referred to as "the multiple bond rule". (iv) The repulsion of electron pairs in filled valence shells is stronger than in incompletely filled valence shells.

The effectiveness and also the beauty of the VSEPR model comes mainly from the fact that the above extremely simple rules may find applications so extensively. On the other hand, it is also important to realize the limitations of this model. It is the knowledge of limitations that makes the application of the model more reliable.

Here it is of interest to inspect some plots of localized orbitals in Figure 32 corresponding to bonds and lone pairs of electrons as a visual demonstration and conformation of the basic assumptions of the VSEPR model. These plots originate from a series of ab initio molecular orbital calculations (Schmiedekamp, Skaarup, Pulay, Hargittai, Cruickshank and Boggs, 1977) using the force method (program MOLPRO; Pulay, 1969). In order to correlate the wavefunctions with the VSEPR model, the localized orbitals were produced by maximizing the sum of the squares of the distances between the centroids of charge of the orbitals (Boys, 1966). As is seen on the plots, the lone pair of electrons occupies more space indeed than bonding pairs around the atom. Also, bonds to more electronegative atoms such as fluorine occupy less space around the central atom than bonds to less electronegative atoms such as hydrogen. The angular range of corresponding contours on the electron density plots are in all cases in excellent agreement with the postulates of the VSEPR model, the difference between lone pairs of electrons and bonding pairs being especially large.

Returning to the discussion of the structural peculiarities of the sulphones, the most striking feature of an XSO_2Y molecule

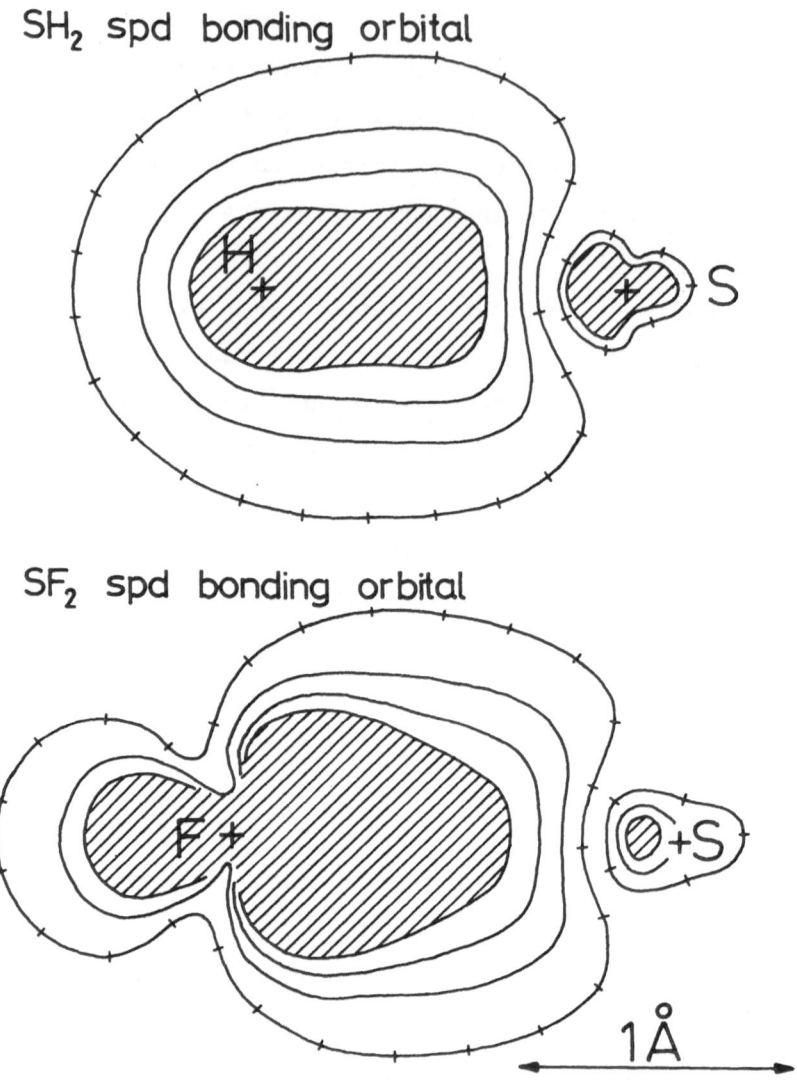

Figure 32

Pictorial representation of localized orbitals (A. Schmiedekamp: Dissertation, Austin, Texas, 1976)

SOH$_2$ spd bonding orbital

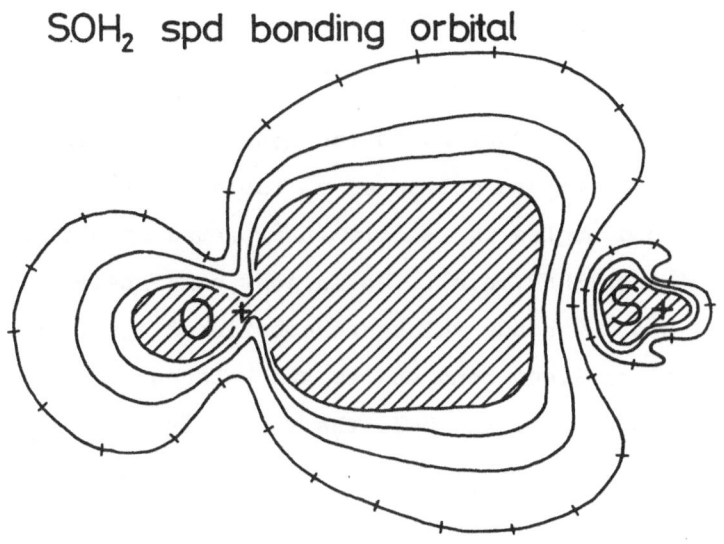

SF$_2$ spd lone pair orbital

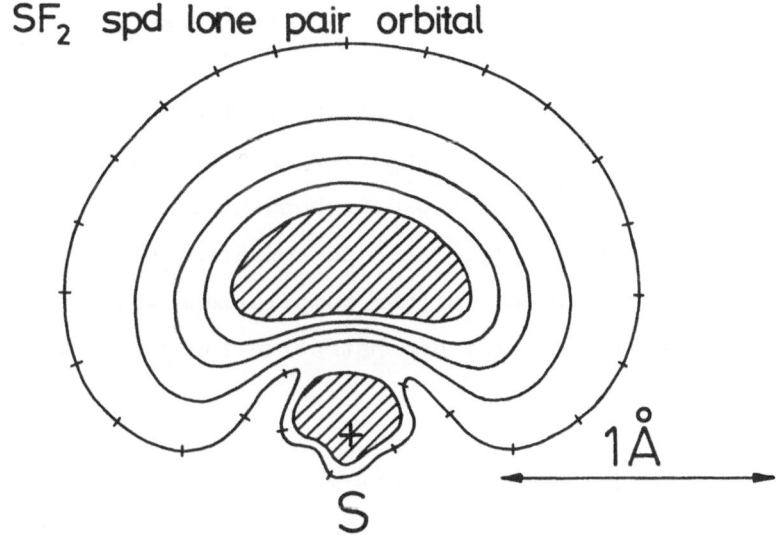

Figure 32 (continued)

The contour lines denote the following electron densities: 0.02, 0.06, etc., electron a_0^{-3} ($a_0 \approx 0.53$ Å)

is the relationship between the bond angles

$$X-S-Y \; \langle \; X(Y)-S=O \; \langle \; O=S=O \; ,$$

and this feature is invariant to the nature of the ligands X and Y. The relevant data are collected in Table 22. This relationship between the bond angles is readily interpreted by the multiple bond rule of the VSEPR model. The largest bond angles appear between the two S=O bonds which are considered essentially double bonds, and the smallest bond angles are between the two essentially single bonds S-X and S-Y. The former are much larger while the latter are considerably smaller than the ideal tetrahedral angle[*].

Consider now other variations in the geometrical parameters from the point of view of the VSEPR model. The parameters are grouped in Tables 23-26. S-C bond lengths, and C-S-X bond angles (X = F, Cl or CH_3) in methane sulphonyl derivatives are collected in Table 23. Sulphur-halogen bond lengths and X-S-F bond angles (X = F, CH_3O or CH_3), and O=S-Cl bond angles are given in Tables 24 and 25, respectively. Table 26 contains the S=O bond lengths and O=S=O bond angles. Further compilations may also be made, but these are sufficient for demonstrating the points thought to be important.

---

[*] Apparently the only exception is N,N'-tetramethyl sulphonyl diamine in which both gas electron diffraction and crystalline X-ray diffraction determined larger than tetrahedral N-S-N bond angles (110.5 ± 0.4 and $112.6\pm0.4^{\circ}$, respectively).

Table 22

Sulphur bond angles in sulphone molecules

Compound	$\angle$ (°)		
	X–S–Y	X(Y)–S=O	O=S=O
FSO_2F	96	108	124
CH_3OSO_2F	97	107, 109	124
CH_3SO_2F	98	106, 110	123
$ClSO_2Cl$	100	108	123
CH_3OSO_2Cl	103	106, 109	122
CH_3SO_2Cl	101	107, 109	121
$CH_3SO_2CH_3$	103	108	120
$CH_2=CHSO_2Cl$	100	106, 110	122
$C_6H_5SO_2Cl$	101	106, 110	122
$OCNSO_2Cl$	98	108, 108	123
$(CH_3)_2NSO_2Cl$	103	106, 109	123

Before inspecting the tendencies in the individual series, two comments are made. First, that although the changes in the geometrical parameters are examined as the ligand electro-negativities change, it is recognized that the atomic groups have no well-defined electronegativity values. Thus only the general tendency is to be considered. Second, that the changes in the geometrical parameters from compound to compound may be commen-surable with or even smaller than the experimental uncertainties.

Table 23

The S-C bond lengths and S-C-X bond angles in

CH_3SO_2X molecules (X = F, Cl, CH_3)

Compound	r(S-C) (Å)	∠ C-S-X (°)
CH_3SO_2F	1.759(6)	98.2(15)
CH_3SO_2Cl	1.763(5)	101.0(14)
$CH_3SO_2CH_3$	1.771(4)	102.6(9)

Table 24

The S-F bond lengths and X-S-F bond angles in

XSO_2F molecules (X = F, CH_3O, CH_3)

Compound	r(S-F) (Å)	∠ X-S-F (°)
FSO_2F	1.530(3)	96.1(2)
CH_3OSO_2F	1.545	96.8
CH_3SO_2F	1.561(4)	98.2(15)

Thus, again, it is important to observe the tendencies in the structural variations in relatively large series of compounds, and the tendencies observed in small series should be considered tentative only.

Keeping in mind the above warning, let us consider now the following tendencies which accompany the increase of the ligand electronegativities:

Table 25

The S-Cl bond lengths, O=S-Cl bond angles, and O...Cl
distances in sulphonyl chloride molecules

Compound	r(S-Cl) (Å)	∠O=S-Cl (°)	r(O...Cl) (Å)
FSO_2Cl	1.985	107.5	(2.757)
$ClSO_2Cl$	2.011	107.7	2.781
$OCNSO_2Cl$	2.018	107.8	2.798
CH_3OSO_2Cl	2.023	106.3	2.779
$CH=CHSO_2Cl$	2.035	106.3	2.789
$C_6H_5SO_2Cl$	2.047	105.5	2.780
CH_3SO_2Cl	2.046	107.1	2.816
$(CH_3)_2NSO_2Cl$	2.064	105.8	2.806

(i) The S-X (X = O, C, F or Cl) bonds shorten;

(ii) The O=S=O bond angles open;

(iii) The C-S-X (X = F, Cl or CH_3) and X-S-F (X = F, CH_3O or CH_3) bond angles decrease;

(iv) No tendency is observed in the changes of the O=S-Cl bond angles.

According to the VSEPR model, the more electronegative ligand draws away electron density from the central sulphur atom, the space requirement of, and accordingly, the repulsion from the bonding pair decrease ("the electronegativity rule"). As the neighbouring bonds suffer less repulsion from the bonding pair

considered, they may shorten. At the same time the bond angle in
whose formation the sulphur-ligand bond participates, may decrease. All this accounts for tendencies (i) and (iii).

According to the multiple bond rule of the VSEPR model, the
shortening of the S=O bonds, i.e. the increase in their multiple
bond character, results in an opening of the O=S=O bond angle,
hence tendency (ii).

Note, however, that according to the electronegativity rule
the presence of a more electronegative ligand attached to the
SO_2 group tends to shorten all the other sulphur bonds. This, in
turn, should open the bond angles according to the multiple bond
rule. This is not observed, however as seen in tendency (iii),
and it would also contradict to the electronegativity rule
indeed. In any case the variations in the C-S-X and X-S-F bond
angles collected in Tables 23 and 24 depend on the relative magnitude of the two effects mentioned, which turn out to be competing.
From point of view of the bond angles given as examples, the
effect described by the electronegativity rule seems to be
prevailing.

Consider now the changes in the O=S-X bond angles with
changing electronegativity of ligand X. There are again two
competing effects according to the VSEPR model. On the other hand,
the changes in the electronegativity of ligand Y in $X-SO_2-Y$
molecules generate effects which exercise influence in the same
direction as regards the O=S-X bond angles whether the electronegativity or the multiple bond rule is considered. Namely,
increasing the electronegativity of ligand Y should result

always in opening the O=S-X bond angle. This is indeed observed
in the variations of the O=S-F bond angles in the series CH_3SO_2F,
CH_3OSO_2F, FSO_2F (106.2°, 106.8°, 108°), and this would be expected
in the variations of the O=S-Cl bond angles. No such tendency is
observed, however, for the latter as seen in Table 25. One possible explanation is, of course, the experimental error. A
closer inspection of the data reveals that the only considerable
(though still not significant) deviation from the expected
tendency is shown by the O=S-Cl bond angle of methane sulphonyl
chloride. Incidentally, this parameter is relatively well determined. It is also noteworthy that the X-S-Cl bond angles of
XSO_2Cl molecules do not follow the variations in the X-S-F bond
angles of the XSO_2F molecules. Here again the two cited rules of
the VSEPR model represent competing effects so the situation is
rather complicated. The suspicion arises, however, that the VSEPR
model may be less applicable for sulphonyl chlorides than for
sulphonyl fluorides. This would be consistent with the general
observation according to which the importance of non-bonded
interactions (or atom-atom interactions) versus the electron pair
repulsions increases as the size of the ligands increases relative
to the size of the central atom.

Attention has been drawn repeatedly to the possible importance of non-bonded interactions in sulphone molecular
geometries, and in particular in the variations of the geometry
of the SO_2 group (Hargittai and Hargittai, 1974; Hargittai, 1974b).
The oxygen...oxygen distance is strikingly constant at around

Table 26

The S=O bond lengths ($\overset{o}{A}$), O=S=O bond angles (^{o}), and O...O distances ($\overset{o}{A}$) in the sulphone molecules[a]

Compound	r(S=O)		∠O=S=O		r(O...O)	
FSO_2F	1.398(2)	ED	125.1[c]		2.481(6)	MW
FSO_2OCH_3	1.410(3)	ED	124.4(7)	ED	2.484(3)	MW
FSO_2Cl	[1.408][b]		(123.7)[d]		2.484(3)	MW
FSO_2Br	[1.407][b]		(123.7)[d]		2.486(3)	MW
FSO_2CH_3	1.411(3)	ED	123.1[c]		2.480(3)	MW
$ClSO_2Cl$	1.405(4)	ED	123.5(10)	ED	2.485(3)	MW
$ClSO_2OCH_3$	1.420(4)	ED	(122.2)[d]		[2.485][b]	
$ClSO_2CH=CH_2$	1.421(6)	ED	122.0(10)	ED	2.484(11)	ED
$ClSO_2C_6H_5$	1.418(12)	ED	122.5(36)	ED	2.482(39)	ED
$ClSO_2NCO$	1.417(3)	ED	(122.8)[d]		[2.484][b]	
$ClSO_2N(CH_3)_2$	1.422(4)	ED	122.7(23)	ED	2.494(26)	ED
$ClSO_2CH_3$	1.425(3)	ED	120.8(8)	ED	2.483(3)	MW
$CH_2=CHSO_2CH=CH_2$	1.440(4)	ED	(119.5)[d]		[2.484][b]	
$(CH_3)_2NSO_2N(CH_3)_2$	1.433(10)	ED	(119.8)[d]		[2.48][b]	
$CH_3SO_2CH_3$	1.436(3)	ED	119.7(11)	ED	2.493(3)	MW
SO_2	1.432(2)	ED			2.476(3)	MW
$(CH_2)_4SO_2$	1.449(10)	ED	115(3)	ED	2.44	ED
$(CH_2)_2SO_2$	1.439(6)	MW	121.4(5)	MW	2.501(3)	MW

[a] For detailed references and comments see preceding tables and text; The ED S=O values are r_g, the MW O...O values are r_0; ED – electron diffraction, MW – microwave spectroscopy;

Table 26 (continued)

The average of the MW O...O distances is 2.485 (σ = 0.007)Å; without the datum on $(CH_2)_2SO_2$, 2.484 (σ = 0.004)Å.

[b] Assumed

[c] Calculated from electron diffraction r(S=O) and microwave spectroscopic r(O...O)

[d] In calculating these bond angles one of the distances was assumed.

2.48-2.49 Å in a large series of molecules[*] as seen on the data of Table 26 and is well demonstrated by Figure 33. This Figure shows the S=O bond lengths (R) and O=S=O bond angles (2A) from electron diffraction investigations, and curves determined by a least-squares procedure on the basis of the experimental data. As the observation on the constancy of the O...O distances is utilized, the relationship between the S=O bond distances and O=S=O bond angles can be found from

$$2A = 2 \text{ arc sin} \frac{2.483}{2R} \quad ,$$

---

[*] This phenomenon has, in fact, been utilized in several structure analyses, e.g. those on chlorosulphuric acid methyl ester (Hargittai, Schultz and Kolonits, 1977), divinyl sulphone (Hargittai, Rozsondai, Nagel, Bulcke, Robinet and Labarre, 1977), sulphonyl chloride isocyanate (Brunvoll, Hargittai and Seip, 1977).

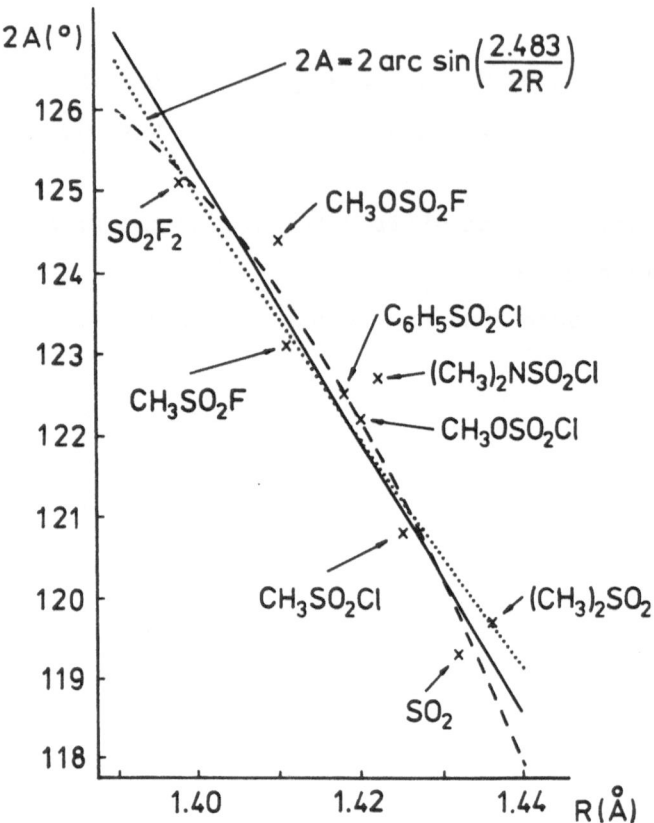

Figure 33

The O=S=O bond angle (2A) as a function of the S=O bond lengths
(R)

+ experimental values (electron diffraction)

—— linear equation (Table 42)

--- second degree equation (Table 42)

... 2A = 2 arc sin (2.483/2R)

and the curve from this function differs very little indeed from
the curves found by the least-squares procedure (see in Table 42;
Brunvoll and Hargittai, 1977b).

The above observation indicates that in addition to the
electron pair repulsions considered in the VSEPR model, the non-
bonded interactions may be equally important in establishing the
actual geometry of the sulphur bond configuration in the sulphone
molecules. It may be visualized that the sulphur atom of an XSO_2Y
molecule is situated in the centre of the tetrahedron with the
two oxygen atoms in two of the four apexes of the tetrahedron.
As the other two ligands (X and Y) are changed, the two oxygen
atoms remain in their position while the sulphur atom may move
up and down along the direction as a continuation of the bisector
of the O=S=O angle.

It certainly seems to be important to consider the internuclear distances between non-bonded atoms, in addition to the
bond lengths and bond angles, when discussing and interpreting
geometrical variations in molecular structures.

The oxygen-chlorine non-bond distances in some sulphonyl
chloride derivatives are given in Table 25. Below, some other
non-bond distances are listed providing further examples of
their constancy (references are not given here, they may be
found according to the formulae in previous sections):

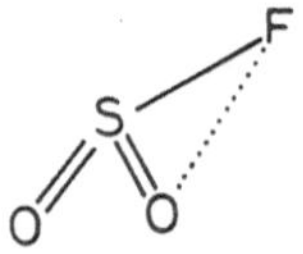

2.38 Å SO$_2$F$_2$

2.37   CH$_3$OSO$_2$F

2.38   CH$_3$SO$_2$F

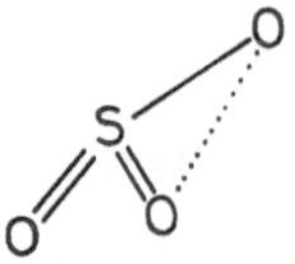

2.42 Å CH$_3$OSO$_2$F

2.42   CH$_3$OSO$_2$Cl

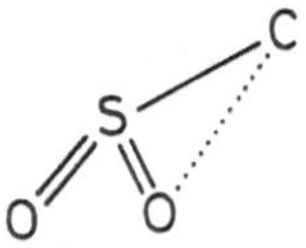

2.61 Å (CH$_3$)$_2$SO$_2$

2.60   CH$_3$SO$_2$F

2.61   CH$_3$SO$_2$Cl

2.61   C$_6$H$_5$SO$_2$Cl

2.60   CH$_2$=CHSO$_2$Cl

Summarizing, it may be stated that both electron pair repulsions (some of them resulting in competing effects), and non-bonded interactions seem to be of importance in establishing the sulphone molecular geometries. The actual structures depend on the relative magnitudes of the various interactions.

Crystal-phase SO$_2$ geometries

Characteristic variations in the S=O bond lengths and O=S=O bond angles were observed in vapour-phase sulphone molecules depending on the nature (electronegativity) of the ligands attached to the SO$_2$ group. Data are very scarce referring to sulphone molecules whose molecular structure has been elucidated

both in vapour and crystalline phases. The relevant data have
been collected (cf. dimethyl sulphone in Table 11, N,N'-tetra-
methyl sulphonyl diamine in Table 19). They do not warrant,
however, a detailed comparison. It is only hoped that the amount
of structural information will gradually grow concerning vapour-
phase and crystal-phase species in order to enable us to examine
finer details of changes upon transition from one phase into
another. Since there are numerous crystal-phase molecular
structure determinations by X-ray diffraction (in fact much more
than vapour-phase studies), it is of considerable interest to see
whether the most characteristic structural variations observed in
free molecules are also present in the crystal-phase structures.

Although a complete coverage was attempted for the vapour-
phase structures, it is stressed that no such aim was pursued in
case of the crystal-phase structures. Indeed, literature data
were collected in a very casual manner and no critical approach
was applied in presenting the parameters from the original papers.
It is also emphasized that it was not examined whether correction
for thermal motion has been applied by the authors although its
importance cannot be overemphasized. This is illustrated by the
uncorrected and corrected parameters of the SO_2 group in
3,4-epoxy sulpholane (Sands, 1972):

$C_4H_6OSO_2$	uncorrected	corrected
$r(S=O)$ (Å)	1.444 ± 0.002	1.455
$\angle O=S=O$ (°)	117.7 ± 0.2	117.8
$r(O...O)$ (Å)	2.470 ± 0.004	2.492

The data presented in Tables 27 and 28 are thus good enough for giving a general impression only. The two tables contain data on organic and inorganic substances. The substances were considered to be inorganic if both bonds of the SO_2 group are linking non-carbon atom. The inorganic sulphones are considered separately because of their structural peculiarities and the greater variety of their parameters. The mean values and standard deviations of the parameters presented for the organic sulphones are also given in Table 27.

Some comments are made concerning the crystal-phase inorganic sulphone structures. In the S_3O_9 molecule the endocyclic O...O distance is 2.46 Å, i.e. almost the same as the exocyclic O...O distance. The latter is, of course, the one to be compared with the analogous parameter of the SO_2 groups of other sulphones. The mean length of the endocyclic S-O bonds is 1.626±0.007 Å and the mean endocyclic sulphur bond angle is 98.7° (McDonald and Cruickshank, 1967).

Some of the non-bond distances of the $FXeOSO_2F$ molecule are given in the Figure illustrating the molecular model (Bartlett, Wechsberg, Jones and Burbank, 1972). These non-bond distances are similar to analogous ones observed in vapour-phase molecules.

The SO_2 parts are considered to be Lewis acid in both the sulphur dioxide trimethyl amine complex (Helm, Childs and Christian, 1969) and the iridium compound (La Placa and Ibers, 1966). The SO_2 group as Lewis acid is supposed to draw electron density away from the nitrogen and iridium atoms, respectively, and this is consistent with the observation of the pyramidal

Table 27

The geometry of SO_2 groups in crystal-phase sulphones (organic)

Compound		r(S=O) (Å)	∠O=S=O (°)	r(O...O) (Å)
1	a	1.445	117.9	2.48
	b	1.446	117.3	2.47
2		1.436	119.2	2.48
3		1.437	117.3	2.45
4		1.435	118.3	2.46
5		1.434	118.5	2.46
6		1.432	118.1	2.46
7		1.440	118.4	2.48
8		1.444	117.7	2.47
9		1.456	117.8	2.49
10		1.435	117.5	2.45
11		1.49	120	2.58
12		1.448	116.5	2.47
13		1.439	116.5	2.45
14		1.433	116.8	2.44
15		1.440	117.0	2.46
16		1.424	117.6	2.44
17		1.428	120.3	2.48
18		1.446	119.7	2.50
19		1.44	119	2.48
19		1.451	118.2	2.49
19		1.445	117.3	2.45
20		1.456	118.3	2.50
21		1.434	118.6	2.47
22		1.432	118.1	2.46
23		1.412	120.6	2.45
24		1.432	120.0	2.48
25	✶	1.424	118.3	2.44
	✶✶	1.417	119.1	2.44
26		1.427	118.9	2.46

Table 27 (continued)

average	1.439	118.3	2.470
σ	0.014	1.1	0.027

1 a	Sands (1963)
1 b	Langs, Silverton and Bright (1970)
2	Sime and Woodhouse (1974)
3	Chawdhury (1976)
4	Harlow, Sammes and Simonsen (1974)
5	Harlow, Simonsen, Pfluger and Sames (1974)
6	Loughry and Simonsen (1975)
7	Wadhawan (1976)
8	Sands (1972)
9	Towns and Simonsen (1974)
10	Towns and Simonsen (1975)
11	Kronfeld and Sass (1968)
12	Jandacek (1968)
13	Jandacek (1968)
14	Guin (1969)
15	Sands and Day (1967)
16	Beall, Herdklotz and Sass (1970)
17	Kiers and Vos (1972)
18	Noordik and Vos (1967)
19	O'Connor and Maslen (1965)
19	O'Connel and Maslen (1967); Alléaume and Decap (1965a)
19	Alléaume and Decap (1965b)
20	Alléaume and Decap (1968)
21	Klug (1968)
22	Kálmán, Duffin and Kucsman (1971)
23	Cotton and Stokely (1970)
24	Karle (1973)
25	Dupont and Dideberg (1972)
26	Gogoi and Hargreaves (1970)

1 $CH_3-\overset{\overset{\displaystyle O}{\|}}{\underset{\underset{\displaystyle O}{\|}}{S}}-CH_3$

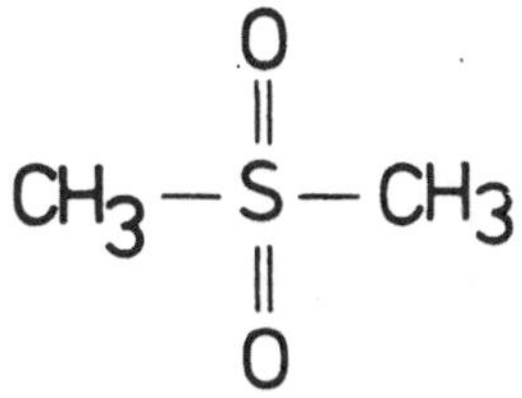

2

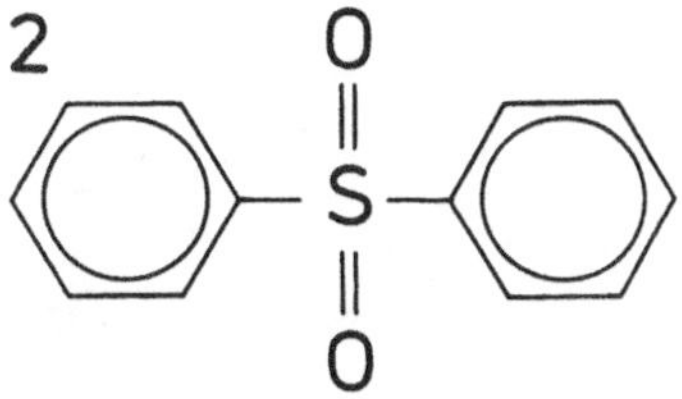

3

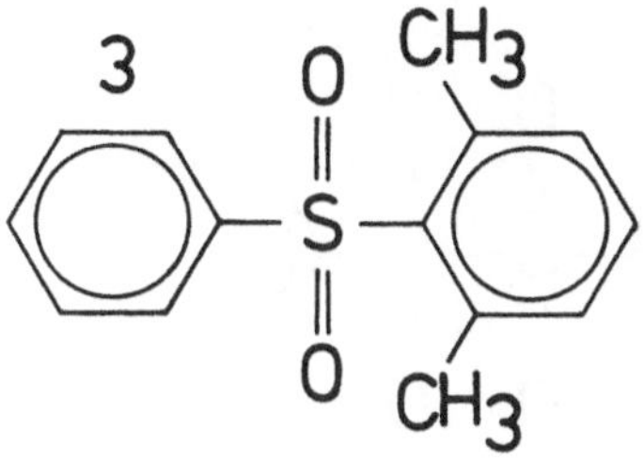

4

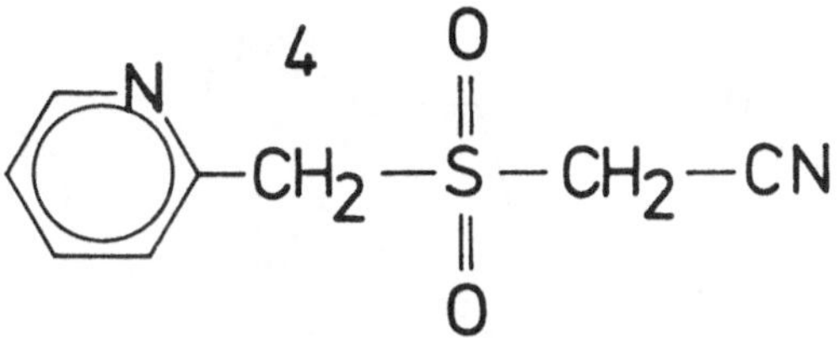

5

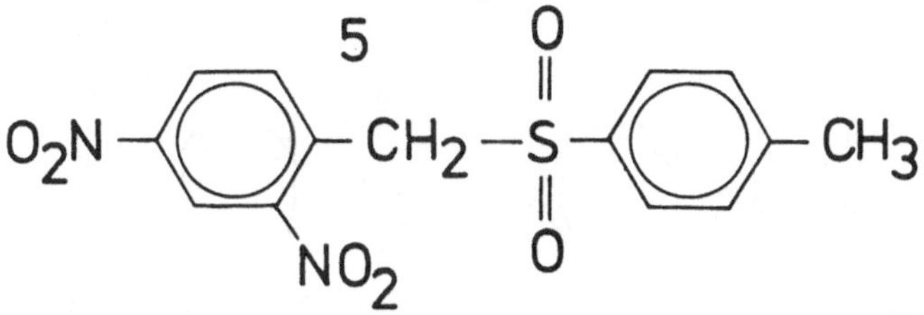

6

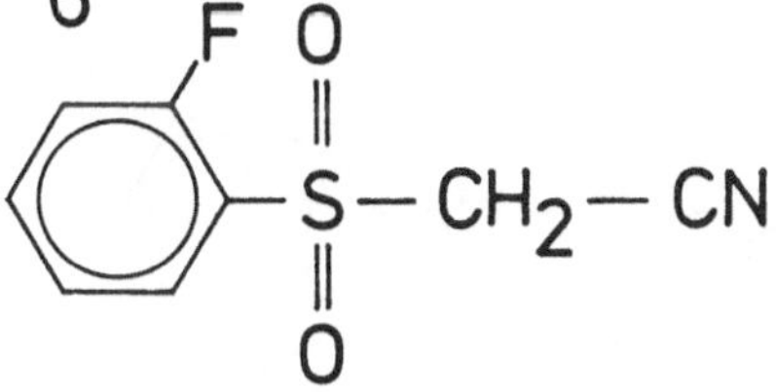

7

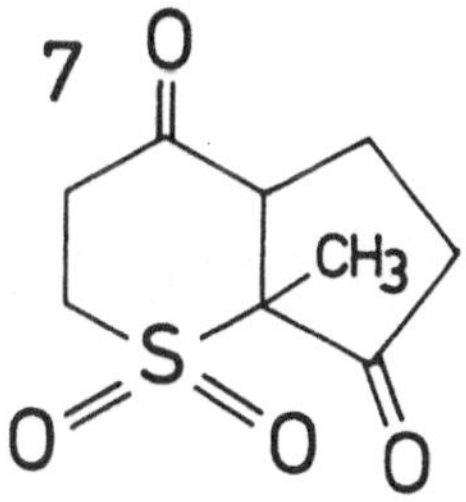

8

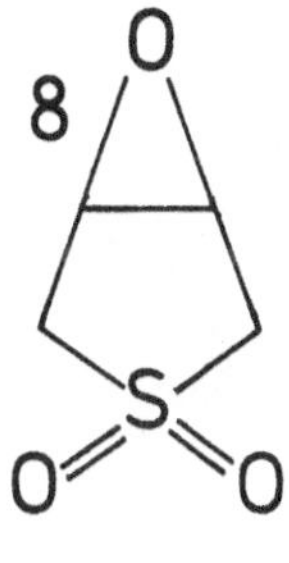

9

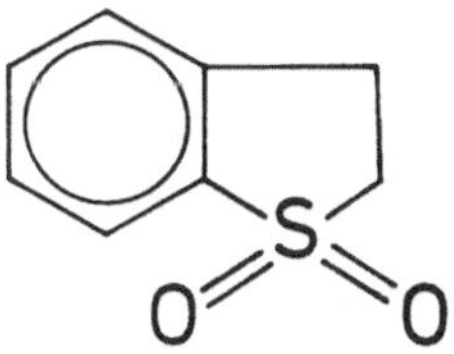

10

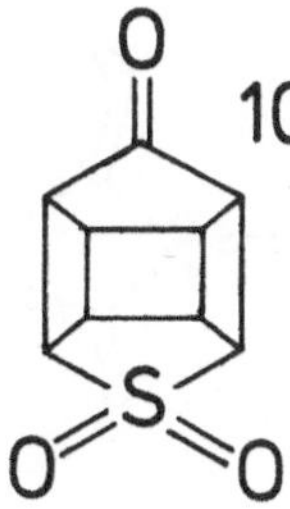

11

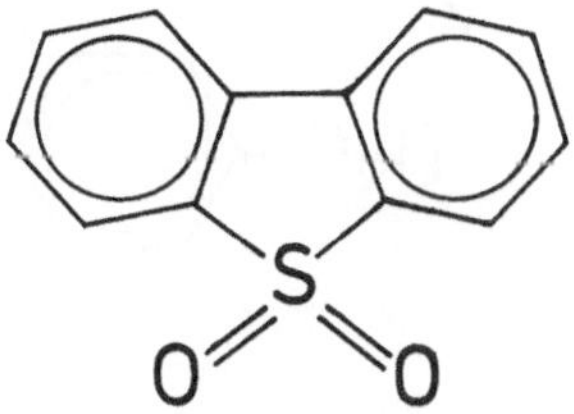

References in Table 27

References in Table 27

References in Table 27

configuration. In order to account for the structure of the
ruthenium compound (Vogt, Katz and Wiblerley, 1965), it was
supposed that the sulphur dioxide part acts as Lewis base do-
nating electron density towards the ruthenium atom, hence the
planar sulphur bond configuration. Thus the bonding peculiarities
in these substances are very different (cf. Shriver, 1970).

Table 28

The geometry of SO_2 groups in crystal-phase sulphones (inorganic)

Compound		r(S=O) (Å)	∠ O=S=O (°)	r(O...O) (Å)
$H_2NSO_2NH_2$	i	1.391(8)	119.4(8)	2.40
$(CH_3)_2NSO_2N(CH_3)_2$	ii	1.449, 1.441(5)	119.7(4)	2.50
S_3O_9	iii	1.428, 1.367(25)	127.9(15)	2.50
$FXeOSO_2F$	iv	1.415, 1.430(9)	119.6(6)	2.47
$[Fe(CO)_2C_5H_5]_2SO_2$	v	1.480, 1.476(1)	112.9(8)	2.46
$(CH_3)_2N \cdot SO_2$	vi	1.377, 1.416(8)	114.8(5)	2.35
$Ir(CO)x$				
$x[P(C_6H_5)_3]_2ClSO_2$	vii	1.411, 1.467(25)	117.1(15)	2.46
$[RuCl(NH_3)_4SO_2]Cl$	viii	1.394, 1.462(10)	113.8(6)	2.39

i   Trueblood and Mayer (1956)

ii   Jordan, Smith, Lohr and Lipscomb (1963)

iii   McDonald and Cruickshank (1967)

iv   Bartlett, Wechsberg, Jones and Burbank (1972)

v   Churchill, Deboer and Kalra (1972)

vi   Helm, Childs and Christian (1969)

vii   La Placa and Ibers (1966)

viii   Vogt, Katz and Wiblerley (1965)

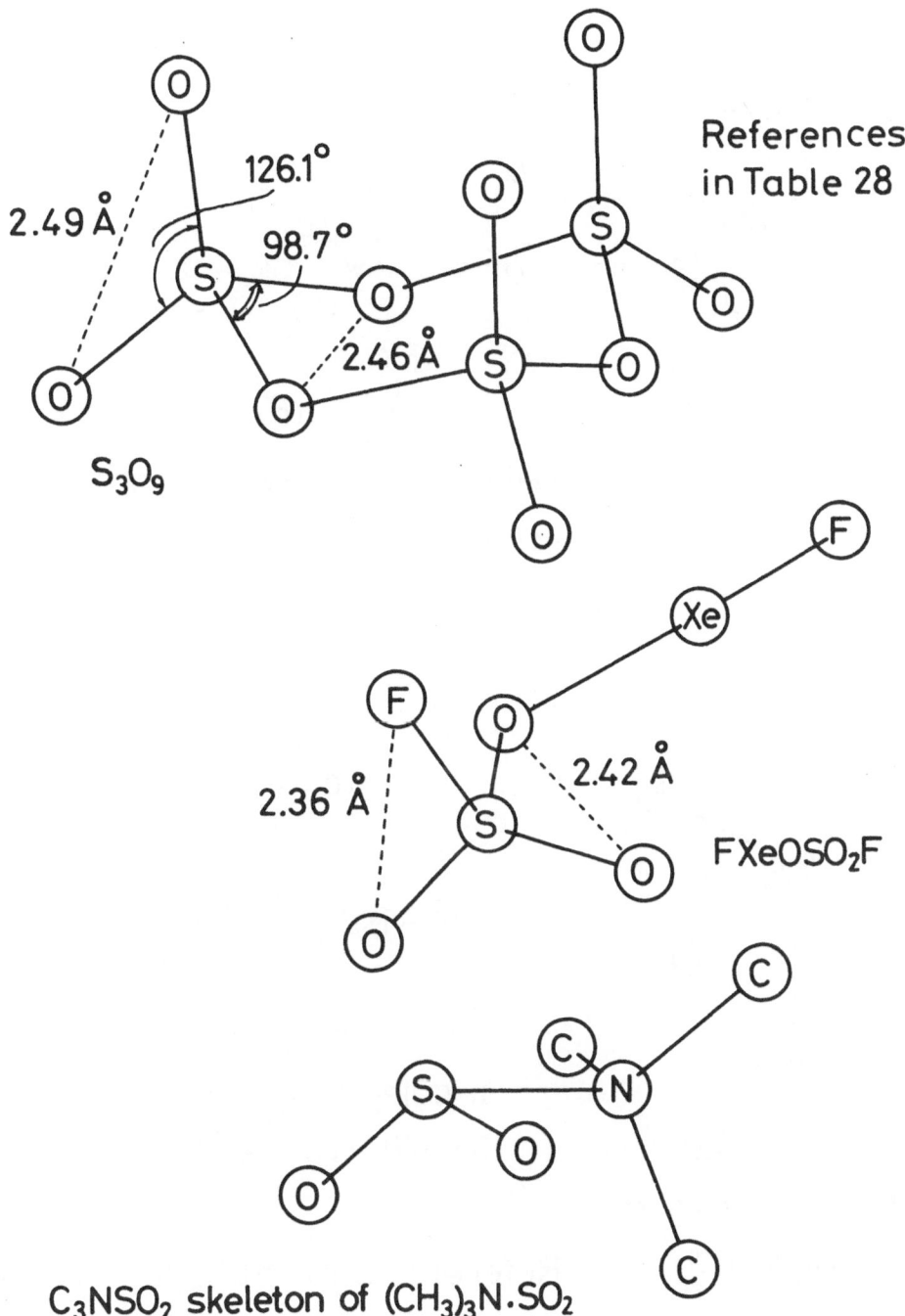

126.1°

2.49 Å

98.7°

2.46 Å

S_3O_9

References in Table 28

2.36 Å

2.42 Å

$FXeOSO_2F$

C_3NSO_2 skeleton of $(CH_3)_3N \cdot SO_2$

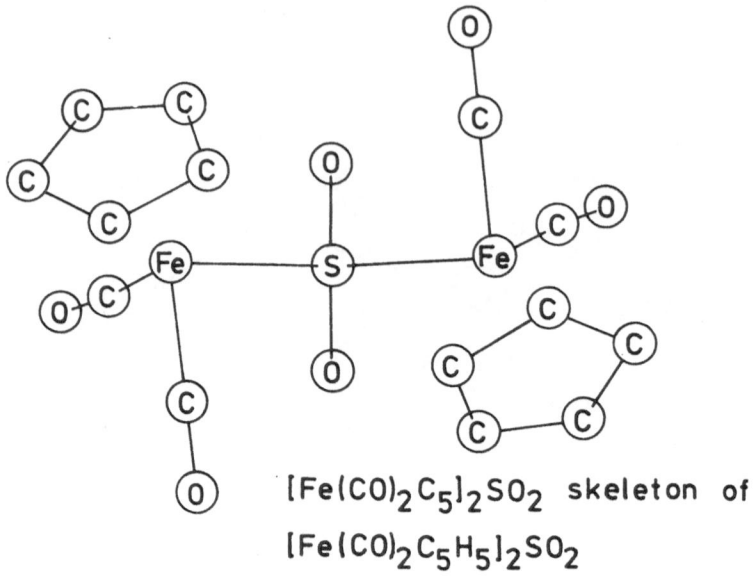

[Fe(CO)$_2$C$_5$]$_2$SO$_2$ skeleton of
[Fe(CO)$_2$C$_5$H$_5$]$_2$SO$_2$

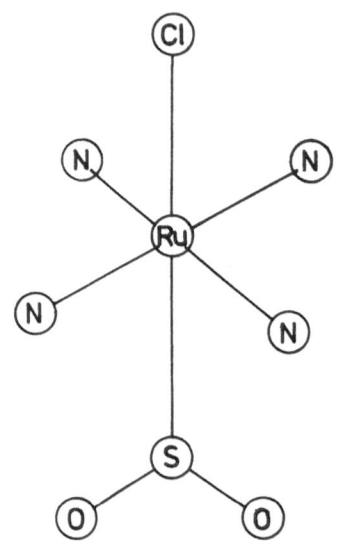

RuClN$_4$SO$_2$ skeleton of
[RuCl(NH$_3$)$_4$SO$_2$]Cl

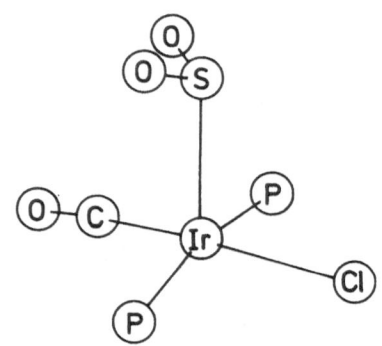

Ir(CO)P$_2$ClSO$_2$ skeleton of
Ir(CO)[P(C$_6$H$_5$)$_3$]$_2$ClSO$_2$

References in Table 28

Comparison of analogous sulphone, sulphoxide, and sulphide geometries

There are four different ligands X for which complete structural comparison is possible in the series SO_2X_2, SOX_2, and SX_2. These ligands are the dimethylamino and methyl groups, and chlorine, and fluorine atoms. The relevant bond angles and bond lengths are compiled in Table 29, and the variations in the bond angles are also depicted in Figure 34 because of their special importance.

As the structural changes are inspected going from the sulphones toward the sulphides, what is seen is indeed the consequence of the bonding pairs replaced by lone pairs of electrons in the series SO_2X_2, SOX_2E, SX_2E_2, where E denotes the lone pair of electrons on the central sulphur atom.

Going from the four ligand case to the three ligand plus one lone electron pair case it is observed that the bond angles decrease and the bonds lengthen. This is simple to interpret by means of the VSEPR model. As the ligand bonding pairs are replaced by a lone pair of electrons, the latter have larger space requirement around the sulphur atom and, accordingly, exercise more repulsion toward the other bonding electron pairs. As a second ligand is replaced by an other lone pair, a further decrease of the X–S–X bond angle and more lengthening of the sulphur bonds could be expected. However, this is not observed. Instead, the bond angles are larger and the bonds are shorter

Table 29

Comparison of bond angles and bond lengths in analogous
sulphones, sulphoxides and sulphides

Molecules

$[(CH_3)_2N]_2SO_2$	a	$[(CH_3)_2N]_2SO$	d	$[(CH_3)_2N]_2S$	h
$(CH_3)_2SO_2$	b	$(CH_3)_2SO$	e	$(CH_3)_2S$	i
SO_2Cl_2	c	$SOCl_2$	f	SCl_2	j(j')
SO_2F_2	c	SOF_2	g(g')	SF_2	k

X-S-X bond angles ($^{\circ}$)

110.5	96.9	114.5
102.6	96.6	99.2
100.0	96.1	102.8(103.0)
96.1(96.7)	92.2(92.8)	98.2

Bond lengths (Å)

S-X	S=O	S-X	S=O	S-X
1.651	1.432	1.693	1.480	1.688
1.771	1.435	1.799	1.485	1.810
2.011	1.404	2.076	1.443	2.014(2.006)
1.530(1.530)	1.405(1.398)	1.583(1.585)	1.420(1.413)	1.592

a, b, c: see Tables 19, 11, 3, respectively; d: Hargittai
and Vilkov (1970);   e: Feder, Dreizler, Rudolph and Typke
(1969); f: Hargittai (1969); g: Hargittai and Mijlhoff (1973);
g': Lucas and Smith (1972); h: Hargittai and Hargittai (1973b);
i: Radnai, Kolonits, Gregory and Hargittai (1975); j: Murray,
Williams and Weatherley (1972); j': Morino, Murata, Ito and
Nakamura (1962); k: Kirchhoff, Johnson and Powell (1973)

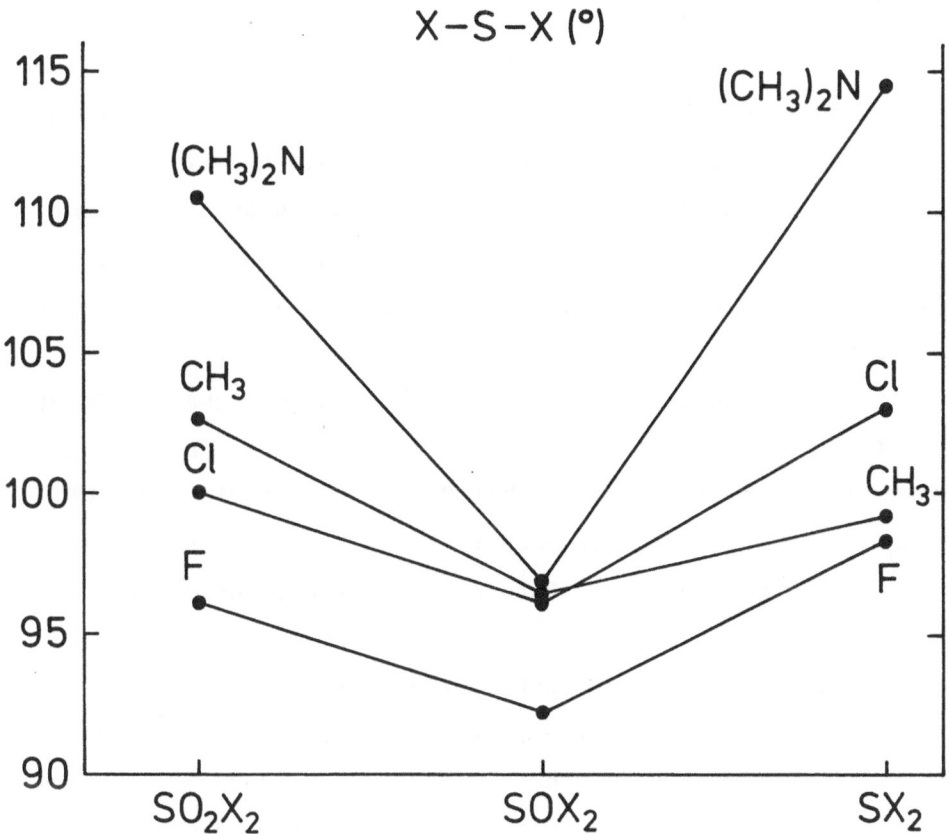

Figure 34

The bond angles X–S–X in a series of sulphones, sulphoxides, and sulphides

in sulphides than in the analogous sulphoxides.

In fact, the application of the VSEPR model for systems with both double bonds and lone electron pairs around the central atom

is not straightforward since the model does not decide whether a
lone pair of electrons or a double bond has more space requirement.
The data for the SO_2X_2 and SOX_2 molecules may indicate that in
case of sulphur as central atom, the lone pair exercises more
repulsion than the double bond. Similar conclusion can be made
for phosphorus as central atom as seen by the bond angles in
Table 30. On the other hand, the data on nitrogen derivatives show
the opposite trend.

It is of interest to examine the geometrical changes in other
AX_4, BX_3E, CX_2E_2 series of molecules in which there are no double
bonds. Gillespie (1972) noted that "... in the series CH_4, NH_3
and H_2O the bond angle decreases from 109.5° to 107.3 and to
104.5° as the number of non-bonding pairs increases." Data on
bond angles for some simple molecules are collected in Table 31
and shown in Figure 35 (Hargittai, 1973). Again, going from the
four ligand case to the three ligand plus one lone electron pair
case is in agreement with the above statement. However, for the
second step, viz. as the second bonding pair is replaced by an
other lone pair, the above statement can be applied for the
molecule pairs NH_3, H_2O and PH_3, SH_2, only. For the other molecule
pairs with chlorine or fluorine ligands, the opposite trend is
realised. Note also the larger than tetrahedral angle in OCl_2
which may be a consequence of the relatively large size of the
ligands.

The interpretation of the changes of the bond angles as
going from the three ligand plus one lone pair case to the two
ligand plus two lone pair case is rather complicated since in

Table 30

Bond angles X-P-X and X-N-X (X = F, Cl, Br or CH_3) in some
simple AOX_3 and AX_3 (A = P or N) molecules

AOX_3		$\angle$X-A-X ($^{\circ}$)	AX_3		$\angle$X-A-X ($^{\circ}$)
POF_3	1	101.3	PF_3	6	97.8
$POCl_3$	1	103.3	PCl_3	7	100.3
$POBr_3$	2	105.6	PBr_3	8	101.0
$PO(CH_3)_3$	3	106	$P(CH_3)_3$	9	98.6
NOF_3	4	100.8	NF_3	10	102.4
$NO(CH_3)_3$	5	109.0	$N(CH_3)_3$	11	110.6

1 Moritani, Kuchitsu and Morino (1971);

2 Olie and Mijlhoff (1969);

3 Wang (1965);

4 Plato, Hartford and Hedberg (1970);

5 Caron, Palenik, Goldish and Donohue (1964);

6 Morino, Kuchitsu and Moritani (1969);

7 Hedberg and Iwasaki (1962);

8 Kuchitsu, Shibata, Yokozeki and Matsumura (1971);

9 Bartell and Brockway (1960);

10 Otake, Matsumura and Morino (1968);

11 Beagley and Hewitt (1968)

Table 31

Bond angles in AX_3E and AX_2E_2 molecules as determined by electron diffraction or microwave spectroscopy

AX_3E		$\angle$ X–A–X ($^{\circ}$)	AX_2E_2		$\angle$ X–A–X ($^{\circ}$)
NH_3	1	108.2	OH_2	7	107.2
NCl_3	2	107.1	OCl_2	8	111.2
NF_3	3	102.4	OF_2	9	103.2
PH_3	4	93.5	SH_2	10	92.2
PCl_3	5	100.3	SCl_2	11	103.0
PF_3	6	97.8	SF_2	12	98.2

1 Kuchitsu, Guillory and Bartell (1968);

2 Bürgi, Stedman and Bartell (1971);

3 Otake, Matsumura and Morino (1968);

4 Bartell and Hirst (1959);

5 Hedberg and Iwasaki (1962);

6 Morino, Kuchitsu and Moritani (1969);

7 Shibata and Bartell (1965);

8 Beagley, Clark and Hewitt (1968);

9 Morino and Saito (1966);

10 Sutton (1965);

11 Morino, Murata, Ito and Nakamura (1962);

12 Kirchhoff, Johnson and Powell (1973)

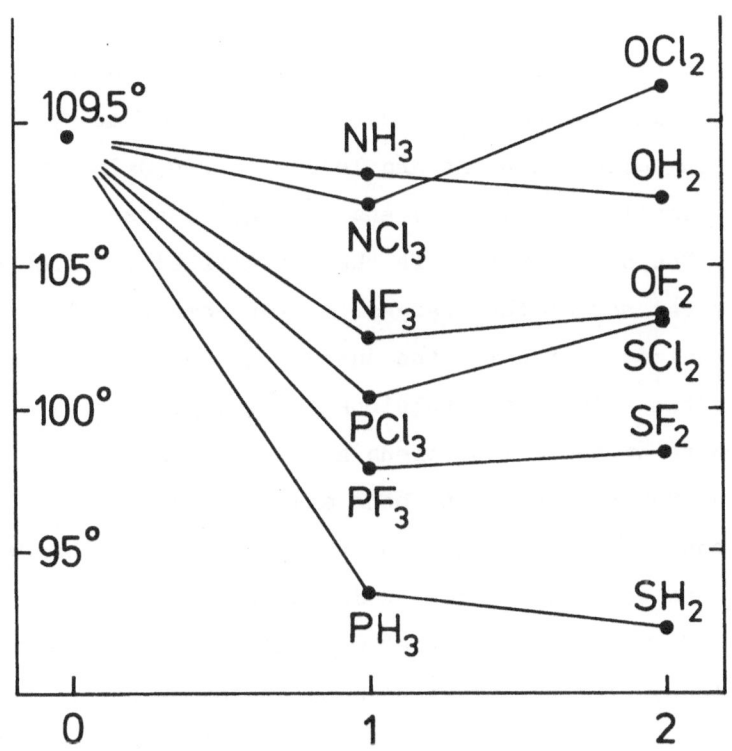

Figure 35

The bond angles X-central atom-X in some series of simple AX_4, BX_3, CX_2 molecules. 0, 1, and 2 indicate the number of lone pairs

addition to the bonding pair - bonding pair and bonding pair - lone pair repulsions, there are also lone pair - lone pair repulsions present. The resulting configuration depends, in the final account, on the relative magnitudes of the three different types of interactions.

To further examine the above changes, a simple point-charges-on-the-sphere model was constructed (Hargittai and Baranyi, 1977) in which bonding electron pairs and lone electron pairs are represented by a smaller charge (q_X) and a larger charge (q_E), respectively. The configuration is then determined in which only radial forces may act on the charges. At the same time it is strongly emphasized that using the charges q_X and q_E implies by no means that the origin of repulsions is simply electrostatic.

The results of the calculations are demonstrated by Figure 36 showing the variations of the bond angles in the AX_3E (Θ) and AX_2E_2 (2β) systems vs. the ratio of the two different charges. The calculations were performed for different values of the repulsion exponent p. It is seen that Θ is always smaller than $109°28'$, while 2β may be larger or smaller than Θ depending on the values of p. Thus, the changes in bond angles as going from AX_4 to AX_3E are generally well understood and are invariant to the choice of the repulsion exponent. On the other hand, the relationship between the bond angles of molecule pairs AX_3E and AX_2E_2 strongly depends on the choice of the repulsion exponents and, accordingly, the simple VSEPR model loses its predictive power for these structural changes.

Ab initio molecular orbital calculations were carried out (Schmiedekamp, Skaarup, Pulay, Hargittai, Cruickshank and Boggs, 1977) in order to gain more insight into the structural variations discussed in this section, and correlate them with bonding properties. Some of the molecules were too large for such

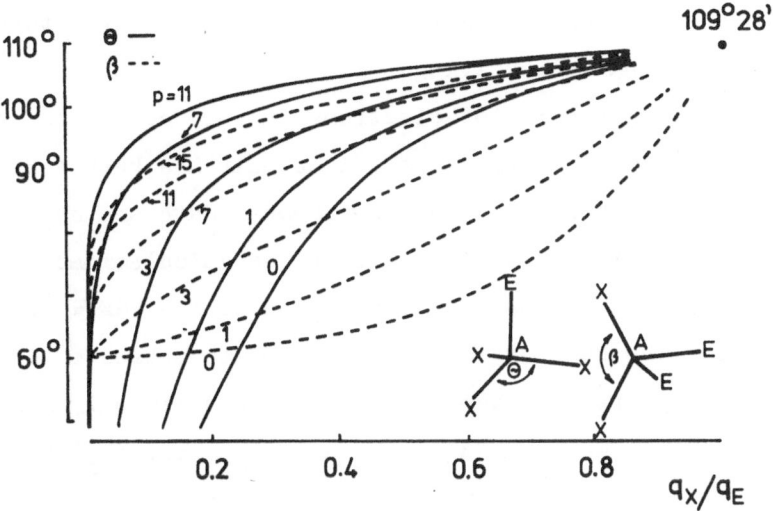

Figure 36

Variations of the bond angles in the AX_3E (Θ) and AX_2E_2 (2β)
systems versus the ratio of the two charges employing various
values for the repulsion exponent in the expression of the
potential

calculations (force method, program MOLPRO; Pulay, 1969),
whereas some other systems, viz. e.g. H_2SO_2, H_2SO, H_2S, seemed
to be of interest to include for comparison. Of course, there
have been calculations for several of the molecules mentioned,
or analogous systems, among others for SO_2H_2, SOH_2 and SH_2
(Van Wazer and Absar, 1972). However, there has been no systematic comparison of these molecules relating wavefunction properties,
obtained with a consistent system of basis sets, and the bond

lengths and bond angles optimized to the lowest energy structure.

The calculated angles involving bonds to hydrogen and fluorine as well as lone pairs of electrons (E) are collected in Table 32. The centroid of charge of the lone pair has been taken as a measure of its position. Where experimental data are available (see preceding Tables), it is seen that the calculations reproduce well the bond angles. The prediction of the VSEPR model according to which the bond angles are expected to decrease as the number of lone pairs increases appears to be valid only for such molecule pairs as NH_3 and NH_2^-, NF_3 and NF_2^-, and SH_3^+ and SH_2, i.e. in cases where the central atom remains the same. On the other hand, the trends observed on the experimental bond angles, i.e. the increase as BX_3E molecules are charged to CX_2E_2, is also seen on the calculated values. It is only the molecule pair NF_3 and OF_2 for which an opposite change is observed in the calculated bond angles. However, the differences are too small to warrant detailed considerations.

An important and unexplained disagreement with the electronegativity rule of the VSEPR model is found in comparing the bond angles in molecule pairs of SH_2 and SF_2, and also PH_3 and PF_3. The experimental and calculated values, however, are consistent with each other.

Analysing the results of the calculations, some interesting observations were made which are discussed in detail by Schmiedekamp, Skaarup, Pulay, Hargittai, Cruickshank and Boggs (1977). They include that d atomic orbitals on the central atom introduce significantly more charge density into the region of

Table 32

Calculated (ab initio, spd basis; parenthesized, sp basis) angles involving bonds and lone pairs of electrons (E) in some simple systems (Schmiedekamp, Skaarup, Pulay, Hargittai, Cruickshank and Boggs, 1977)

Angles (°)	NH_4^+	NH_3E	NH_2E^-	NF_3	NF_2E^-
∠H(F)-N-H(F)	109.5(109.5)	105.4(111.4)	96.7(97.7)	(102.4)	99.4(99.1)
∠H(F)-N-E		113.3(107.5)	107.3(107.4)	(115.9)	103.3(102.9)
∠E-N-E			126.8(125.8)		138.3(139.7)

Angles (°)	SH_3E^+	SH_2E_2	$SHFE_2$	SF_2E_2
∠H(F)-S-H(F)	95.5(99.6)	93.6(95.6)	96.3(95.6)	98.3(98.1)
∠H-S-E	121.3(118.1)	107.8(108.2)	107.6(108.5)	
∠F-S-E			105.1(104.8)	104.4(104.3)
∠E-S-E		126.9(124.5)	130.0(129.2)	135.2(135.8)

Table 32 (continued)

Angles (^{O})	OH_2E_2	OF_2E_2	
$\angle$ H(F)-O-H(F)	104.9(108.1)	102.0(101.7)	
$\angle$ H(F)-O-E	108.2(108.2)	104.3(104.0)	
$\angle$ E-O-E	118.5(115.9)	133.8(135.0)	

Angles (^{O})	SO_2H_2	SOH_2E	SOF_2E
$\angle$ H(F)-S-H(F)	97.5(98.1)	89.2(92.8)	(91.7)
$\angle$ H(F)-S=O	108.3(107.7)	109.9(107.7)	(107.6)
$\angle$ O=S=O	123.2(124.6)		
$\angle$ H(F)-S-E		110.9(111.0)	(109.5)
$\angle$ O=S-E		122.5(122.5)	(125.8)

Angles (^{O})	PH_3E	PF_3E
$\angle$ H(F)-P-H(F)	94.0(95.4)	(96.9)
$\angle$ H(F)-P-E	122.4(121.3)	(120.2)

the bonds taking charge from the central atom, especially from the lone pairs. There is a remarkable constancy of the angular require- ments of a given bonding or lone pair orbital indicating the importance of orbital interactions. The average angular space required by an S-F bond is nearly equal to that of an S-H bond (around 102° and 104°, respectively). The angular requirements of

S=O double bond and a lone pair are almost identical ($113-114°$ and $114-115°$, respectively).

Instructive, however, as the molecular orbital calculations mentioned above are, the bond angle variations in the sulphone, sulphoxide, sulphide series still cannot be considered as well understood.

Geometrical variations in the rest of the molecule

The examination of the changes of the bond angles and bond lengths in the rest of the sulphone molecule may reveal the influence of the sulphur bonding system. The possible interpretation, however, would lead too far away, and we restrict ourselves to compiling and comparing some selected features in related molecules.

The bond angles at two-coordinated oxygen atoms have been determined in several sulphone molecules (cf. Tables 6, 7). They are all considerably larger than the C-O-C bond angle in dimethyl ether ($111.5 \pm 1.5°$; Kimura and Kubo, 1959). Indeed, the bond angles of two-coordinated oxygen atoms show a great variety covering the range between $103°$ to $155°$ (F_2O, see Table 31, $F_3SiOSiF_3$, see Table 33). There seem to be some interesting regularities in the variations of these bond angles. Consider the data collected in Table 33. The X-O-X angles are all larger than the corresponding

Table 33

Bond angles and some bond lengths at the oxygen atom in boron, silicon, phosphorus and sulphur derivatives

∠B–O–C	∠B–O–B	∠Si–O–C	∠Si–O–Si
		F_3SiOCH_3 (1)	$F_3SiOSiF_3$ (2)
		$131.4\pm3.2°$	$155\pm2°$
			△ = 24°
$B(OCH_3)_3$ (7)	$[(CH_3)_2B]_2O$ (8)	$(CH_3)_3SiOCH_3$ (9)	$[(CH_3)_3Si]_2O$ (11)
$121.4\pm0.5°$	$144.4\pm2.7°$	$122.5\pm0.6°$	$148\pm3°$
		$CH_3Si(OCH_3)_3$ (10)	
		$123.7\pm1.0°$	
	△ = 23°		△ = 25°

∠P–O–C	∠P–O–P	∠S–O–C	∠S–O–S
F_2POCH_3 (3)	F_2POPF_2 (4)	FO_2SOCH_3 (5)	FO_2SOSO_2F (6)
$123.7\pm0.7°$	$135.2\pm0.9°$	$116.5\pm0.7°$	$123.6\pm1.2°$
	△ = 11°		△ = 7°
	r(P–O) =		r(S–O) =
=1.560±0.020Å	=1.631±0.005Å	=1.558±0.007Å	=1.611±0.005Å
$P(OCH_3)_3$ (12)			
$119.2\pm1.1°$			
Cl_2SPOCH_3 (13)			
$114.4\pm2.4°$			

Table 33 (continued)

(1) Airey, Glidewell, Robiette and Sheldrick (1971a)
(2) Airey, Glidewell, Rankin, Robiette, Sheldrick and Cruickshank (1970)
(3) Codding, Jones and Schwendeman (1974)
(4) Vilkov and Khaikin (1975) as cited from L.S. Bartell
(5) Hargittai, Seip, Nair, Britt, Boggs and Cyvin (1977)
(6) Hencher and Bauer (1973)
(7) Gundersen (1976)
(8) Gundersen and Vahrenkamp (1976)
(9) Csákvári, Wagner, Gömöry, Hargittai, Rozsondai and Mijlhoff (1976)
(10) Gergő, Hargittai and Schultz (1976)
(11) Csákvári, Wagner, Gömöry, Mijlhoff, Rozsondai and Hargittai (1976)
(12) Vilkov and Khaikin (1975) as cited from H. Oberhammer
(13) Bezzubov and Naumov (1976)

X-O-C angles, X being B, Si, P or S. The difference is the largest for the boron and silicon derivatives and gradually decreases toward the phosphorus and sulphur derivatives. The absolute magnitude of the bond angles shows the same tendency. For the silicon and phosphorus derivatives comparison is possible between molecules in which fluorine atoms or methyl groups are the ligands of the X atom. The bond angles are considerably larger in the fluorine derivatives. If it is true that the presence of the electronegative ligands contracts the d orbitals of the X atom and thus makes them more easily available for bonding and if the consequence of this is the increase in the oxygen bond angle, then the relatively small S-O-C and S-O-S bond angles may be considered as indication of the

sulphur d orbitals participating relatively little in this bonding
system. Our final note in connection with these molecules concerns
the contraversy developed about the structure of F_2POPF_2 . In
addition to the results cited in Table 33, an earlier investigation (Arnold and Rankin, 1972/1973) reported considerably different
results. Vilkov and Khaikin (1975) discussed the differences in
detail. Their conclusions, according to which the data cited in our
Table are preferred, are strengthened by the analogy of the
structural results for the sulphur molecules. For example, the
P–O bonds are expected to lengthen in F_2POPF_2 as compared with
F_2POCH_3 similarly to the changes in the S–O bond lengths[*] (see
Table 33).

The presence of ligands containing silicon, phosphorus,
sulphur, or chlorine influences the nitrogen bond configuration
in decreasing order. This is illustrated by the data collected on
the mean values of nitrogen bond angles (α_N) shown in Table 34.
The averages of the α_N values decrease as follows, 119.5^o, 118.5^o,
115.8^o, and 106.9^o for the silicon, phosphorus, sulphur and
chlorine derivatives, respectively. The sulphones seem to exercise similar influence to that of sulphoxide or sulphide.

---

[*] Arnold and Rankin (1972/1973) reported r(P–O) = 1.533 ± 0.006 Å
along with $\angle$ P–O–P = 145.1 ± 1.2^o from an analysis based on one
conformer present. The later study as cited from L.S. Bartell
by Vilkov and Khaikin (1975) showed the experimental data
to be consistent with a mixture of conformers.

Table 34

The mean values of the nitrogen bond angles and the
lengths of X-N bonds, X = Si, P, S or Cl

Compound		α_N (°)	r(Si-N) (Å)
$(H_3Si)_3N$	(i)	119.7	1.734(2)
$(H_3Si)_2NMe$	(ii)	120.0	1.721(3)
$(H_3Si)_2NN(SiH_3)_2$	(iii)	120.0	1.731(4)
$[(Me_3Si)_2N]_2Be$	(iv)	120.0	1.722(7)
H_3SiNMe_2	(v)	117.0	1.715(4)
F_3SiNMe_2	(vi)	120	1.654(15)
Cl_3SiNMe_2	(vi)	119.8	1.657(12)
$ClSi(NMe_2)_3$	(vii)	119.8	1.715(4)

Compound		α_N (°)	r(P-N) (Å)
$P(NMe_2)_3$	(viii)	117.5	1.700(5)
$ClP(NMe_2)_2$	(ix)	120.0	1.730(5)
Cl_2PNMe_2	(x)	120	1.69(3)
Cl_2OPNMe_2	(x)	116	1.67(4)
F_2PNH_2	(xi)	120.0	1.650(4)
F_2PNMe_2	(xii)	120	1.66
	(xiii)	116.1	1.684(8)

.Table 34 (continued)

Compound		α_N (°)	r(S-N) (Å)
$S(NMe_2)_2$	(xiv)	117.4	1.688(6)
$SO(NMe_2)_2$	(xv)	115.4	1.693(4)
$SO_2(NMe_2)_2$	(xvi)	116.1	1.651(3)
SO_2ClNMe_2	(xvii)	115.4	1.618(5)

Compound		α_N (°)	r(Cl-N) (Å)
Cl_3N	(xviii)	107.1	1.759(2)
Cl_2NMe	(xix)	108.7	1.74(2)
$ClNH_2$	(xx)	104.8	1.7522(1)

(i)   Beagley and Conrad (1970); Hedberg (1955)
(ii)   Glidewell, Rankin, Robiette and Sheldrick (1969)
(iii)   Glidewell, Rankin, Robiette and Sheldrick (1970a)
(iv)   Clark and Haaland (1970)
(v)   Glidewell, Rankin, Robiette and Sheldrick (1970b)
(vi)   Airey, Glidewell, Robiette, Sheldrick and Freeman (1971)
(vii)   Vilkov and Tarasenko (1969)
(viii) Vilkov, Khaikin and Evdokimov (1972)
(ix)   Zaripov, Naumov and Tuzova (1974)
(x)   Vilkov and Khaikin (1966)
(xi)   Brittain, Smith, Lee, Cohn and Schwendeman (1971)
(xii)   Forti, Damiani and Favero (1973)
(xiii) Naumov, Gulyaeva and Pudovick (1972)
(xiv)   Hargittai and Hargittai (1973b)
(xv)   Hargittai and Vilkov (1970)
(xvi)   Hargittai, Vajda and Szőke (1973)

Table 34 (continued)

(xvii)   Hargittai and Brunvoll (1976)
(xviii) Bürgi, Stedman and Bartell (1971)
(xix)   Stevenson and Schomaker (1940)
(xx)   Cazzoli, Lister and Favero (1972)

Table 35 contains X-N=C bond angles and X-N= bond lengths
of isocyanates and isothiocyanates, where X = Si, P, S or Cl.
The large differences in the Si-N=C bond angles reported in
different papers may be due to the different averaging process
over intramolecular motion as these systems were investigated
by microwave spectroscopy and electron diffraction. The bond
angles gradually decrease as going from silicon to chlorine.
Thus the general trend in the bond angle variations is similar
in the systems where atom X is linked to two-coordinated and
three-coordinated (see above) nitrogen atoms.

Correlation seems to emerge between the variations of the
nitrogen bond angles and the X-N bond lengths. As the bond angles
decrease from 120° (three-coordinated nitrogen) and from 180°
(two-coordinated nitrogen), there seems to be a parallel decrease
in the double bond character of the X-N bonds. The diminishing
double bond character of the X-N bonds is demonstrated by the
decrease of the relative shortening of these bonds as compared
with the estimated single bond lengths as seen in Table 36. This
comparison is very crude since there are not enough data to
group them according to the electronegativities of the ligands
attached to atom X. The silicon derivatives show a shortening of

Table 35

Bond angles X–N=C and X–N bond lengths in vapour-phase
isocyanates and isothiocyanates

Compound		∠ X–N=C (°)	r(X–N) (Å)
SiH₃NCO	(a)	180	1.699
	(b)	152.2±1.2	1.703±0.004
SiH₃NCS	(a)	180	1.714±0.010
	(b)	163.8±2.6	1.704±0.006
Si(NCO)₄	(c)	146.4	1.688±0.003
ClSi(NCO)₃	(d)	145±2	1.684±0.003
Cl₂Si(NCO)₂	(d)	136±1	1.687±0.004
Cl₃SiNCO	(d)	138.0±0.4	1.646±0.008
F₃SiNCO	(e)	160.7±1.2	1.648±0.010
F₂PNCO	(f)	130.6±0.8	1.683±0.006
F₂PNCS	(f)	140.5±0.7	1.686±0.007
Cl₂OPNCO	(g)	120±1.5	1.684±0.010
ClO₂SNCO	(h)	122.4	1.656
ClNCO	(i)	118.8±0.5	1.705±0.005

(a) Gerry, Thompson and Sugden (1966)
(b) Glidewell, Robiette and Sheldrick (1972)
(c) Hjortaas (1967)
(d) Hilderbrandt and Bauer (1969)
(e) Airey, Glidewell, Robiette and Sheldrick (1971b)
(f) Rankin and Cyvin (1972)
(g) Naumov, Semashko and Shatrukov (1973)
(h) Brunvoll, Hargittai and Seip (1977)
(i) Hocking and Gerry (1975)

Table 36

Comparison of observed (exp.) and estimated (calc.)
X-N bond distances in two-coordinated and three-
coordinated nitrogen derivatives (X = Si, P, S or Cl)

	exp. ($\AA$)[a]	calc. ($\AA$)[b]	$\triangle$($\AA$)
	r(X-N=)		
Si	1.69	1.82	0.13
P	1.68	1.76	0.08
S	1.66	1.72	0.06
Cl	1.70	1.69	0.01
	r(X-N⟨)		
Si	1.71	1.82	0.11
P	1.63	1.76	0.08
S	1.66	1.72	0.06
Cl	1.75	1.69	-0.06

[a] For two-coordinated nitrogen derivatives cf.
   Table 35
   For three-coordinated nitrogen derivatives cf.
   Table 34

[b] Using the equation of Schomaker and Stevenson (1941)
   $r(X-N) = r_X + r_N - c|\chi_X - \chi_N|$, with c = 0.04 and
   the values of covalent radii (r) and electro-
   negativities (χ) taken from Pauling (1960)

the X-N bond with increasing ligand electronegativity in
agreement with what would be expected.

The N=C bond is longer than the C=O bond in most isocyanates

studied to date in the vapour phase. The relevant data are
collected in Table 37. This supports one of the two models found
for sulphonyl chloride isocyanate (cf. Table 20). However, most
data originate from electron diffraction analyses and the
parameters characterizing these two bonds strongly correlate. It
is suspected that double solutions could have been obtained in
more systems if tested. On the other hand, the N=C bond lengths
reported for isocyanates are similar to those determined in
isothiocyanates. This is reassuring since this bond length can be
determined more reliably in isothiocyanates than in isocyanates
by electron diffraction. The S=C bonds are around 1.56 Å (see
Hargittai and Paul, 1977), thus there is no other bond of similar
length to that of N=C in the thio derivative. In the light of
this the structure of $Cl(O_2)SNCO$ in which $r(N=C) > r(C=O)$ is to
be favoured. At the same time the results reported for CH_3NCO
appear to be suspect.

Table 38 contains lengths of vinyl C=C bonds in a series of
derivatives. The C=C bonds of divinyl sulphone are of the same
length, within experimental error, as that in ethylene. The C=C
bond of vinyl sulphonyl chloride is among the longest, but
unfortunately poorly determined. As noted by Yokozeki and Bauer
(1975) the C=C bonds in substituted derivatives are longer than
in ethylene, except in fluorinated compounds. The exception from
the exceptions is $(CF_3)_2C=CH_2$.

The phenyl carbon-carbon bond lengths of monosubstituted
benzene derivatives and benzene itself are compared in Table 39.
The phenyl ring structure seems to be relatively insensitive to

Table 37

The lengths of N=C and C=O bonds in isocyanates and
isothiocyanates in the vapour phase

Compound		$r(N=C)$ (Å)	$r(C=O)$ (Å)
HNCO	(1)	1.207±0.01	1.171±0.01
CH_3NCO	(2)	1.168±0.005	1.202±0.005
SiH_3NCO		1.216±0.009	1.164±0.008
	(3)	1.200±0.005	[1.180]
	(4)	1.150	1.179
$(CH_3)_3SiNCO$	(5)	1.20±0.01	1.18±0.01
F_3SiNCO	(6)	[1.190]	1.168±0.025
Cl_3SiNCO	(7)	1.219±0.007	1.139±0.008
$Cl_2Si(NCO)_2$	(7)	1.217±0.005	1.146±0.005
$ClSi(NCO)_3$	(7)	1.213±0.005	1.144±0.005
$Si(NCO)_4$	(8)	1.209±0.002	1.165±0.002
GeH_3NCO	(9)	1.190±0.007	1.182±0.007
F_2PNCO	(10)	1.256±0.006	1.165±0.006
$Cl_2(O)PNCO$	(11)	1.161±0.015	1.221±0.015
$Cl(O_2)SNCO$		1.221	1.159
	(12)	1.152	1.218
ClNCO	(13)	1.226±0.005	1.162±0.005
HNCS	(14)	1.216±0.002	
CH_3NCS	(2)	1.192±0.006	
SiH_3NCS	(3)	1.197±0.007	
	(15)	1.211±0.010	
$(CH_3)_3SiNCS$	(5)	1.18±0.01	
F_2PNCS	(10)	1.221±0.006	

(1) Jones, Shoolery, Shulman and Yost (1950)
(2) Anderson, Rankin and Robertson (1972)
(3) Glidewell, Robiette and Sheldrick (1972)

Table 37 (continued)

(4) Gerry, Thompson and Sugden (1966)

(5) Kimura, Katada and Bauer (1966)

(6) Airey, Glidewell, Robiette and Sheldrick (1971b)

(7) Hilderbrandt and Bauer (1969)

(8) Hjortaas (1967)

(9) Murdoch, Rankin and Beagley (1976)

(10) Rankin and Cyvin (1972)

(11) Naumov, Semashko and Shatrukov (1973)

(12) Brunvoll, Hargittai and Seip (1977)

(13) Hocking and Gerry (1975)

(14) Dousmanis, Sanders, Townes and Zeiger (1953)

(15) Jenkins, Kewley and Sugden (1962)

monosubstitution. This applies more to the bond lengths than to the bond angles for which characteristic variations have been observed (Domenicano, Mazzeo and Vaciago, 1976; Domenicano, Vaciago and Coulson, 1975).

In conclusion the remarkably long C-C bond found in thiirane 1,1-dioxide is mentioned (Nakano, Saito and Morino, 1970). It is interesting to compare it with the C-C bonds in the sulphoxide (Saito, 1969) and sulphide (Cunningham, Boyd, Myers and Gwinn, 1951) analogs

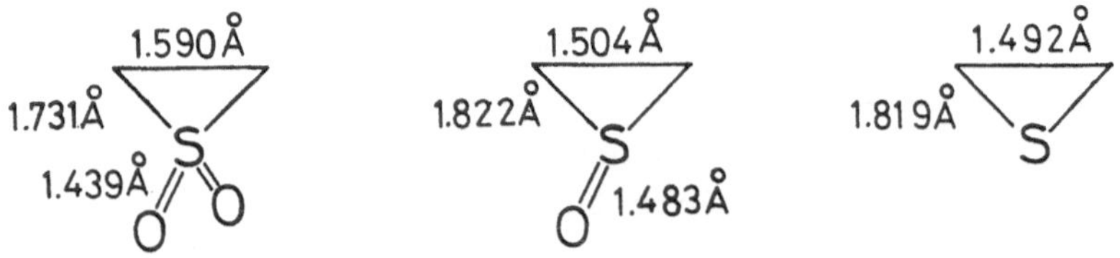

Table 38
The lengths of C=C bonds in ethylene and some
substituted derivatives (vapour-phase data)

Compound		r_g(C=C) (Å)
$H_2C=CH_2$	(1)	1.337 ± 0.002
$F_2C=CH_2$	(2)	1.316 ± 0.006
$FHC=CF_2$	(2)	1.309 ± 0.006
$F_2C=CF_2$	(2)	1.311 ± 0.007
$(CF_3)_2C=CH_2$	(3)	1.374 ± 0.013
$H_2C=CHSCH_3$	(4)	1.342 ± 0.003
$H_2C=C(SCH_3)_2$	(5)	1.349 ± 0.005
$H_2C=C(CH_3)_2$	(6)	1.342 ± 0.003
$(CH_3)_2C=C(CH_3)_2$	(7)	1.353 ± 0.004
$CH_3HC=CHCH_3$ cis	(8)	1.347 ± 0.003
trans	(8)	1.348 ± 0.003
$H_2C=CH-POCl_2$	(9)	1.339
$(CH_2=CH)_2POCl$	(9)	1.346 ± 0.009
$H_2C=CH-SO_2Cl$	(10)	1.357 ± 0.018
$(CH_2=CH)_2SO_2$	(11)	1.333 ± 0.004

(1) Kuchitsu (1966)
(2) Carlos, Karl and Bauer (1974)
(3) Hilderbrandt, Andreassen and Bauer (1970)
(4) Samdal and Seip (1975)
(5) Jandal, Seip and Torgrimsen (1976)
(6) Bartell and Bonham (1960)
(7) Tokue, Fukuyama and Kuchitsu (1974);
    Eisma, Altona, Geise, Mijlhoff and

Table 38 (continued)

Renès (1974); Carlos and Bauer (1974)
(8) Almenningen, Anfinsen and Haaland (1970)
(9) Naumov and Shaidulin (1976)
(10) Brunvoll and Hargittai (1977a)
(11) Hargittai, Rozsondai, Nagel, Bulcke,
    Robinet and Labarre (1977)

Hoffmann, Fujimoto, Swenson and Wan (1973) carried out
semiempirical molecular orbital calculations on this series of
molecules and interpreted the structural variations without
invoking and invoking 3d orbitals of sulphur.

Correlation between geometric and vibrational parameters of the SO_2 groups

Empirical correlations between geometric and vibrational
parameters of relatively simple sulphone molecules have been
established for a number of years (Gillespie and Robinson, 1963)
and used extensively (see, e.g. Banister, Moore and Padley,
1968; Szmant, 1971). At the time when Gillespie and Robinson
(1963) communicated their relationships, it was noted that
relatively few S=O bond lengths were available and they were
considered in most cases to be less accurate than the corresponding frequencies. The usefulness of empirical relationships

Table 39

The mean C-C bond lengths in benzene and
monosubstituted derivatives

Compound		$r(C \dot{=} C)$ (Å)			
C_6H_6	(1)	1.396	r_α^o	; 1.397	r_z
$C_6H_5CH_3$	(2)	1.396	r_α		
C_6H_5CN	(3)	1.395	r_z		
$C_6H_5NH_2$	(4)	1.396	r_s		
C_6H_5OH	(5)	1.398	r_o		
C_6H_5F	(6)	1.392	r_s		
$C_6H_5SO_2Cl$	(7)	1.404(10)	r_g		
C_6H_5Cl	(8)	1.396	r_s		

(1) Tamagawa, Iijima and Kimura (1976)

(2) Seip, Schultz, Hargittai and Szabó (1977)

(3) Casado, Nygaard and Sørensen (1971)

(4) Lister, Tyler, Høg and Larsen (1974)

(5) Pedersen (1969)

(6) Nygaard, Bojesen, Pedersen and Rastrup-Andersen (1968)

(7) Brunvoll and Hargittai (1976)

(8) Michel, Nery, Nosberger and Roussy (1976)

is increasingly recognized which is also shown by recent studies
in this area applying up-to-date experimental information. Thus
for example relationships have been communicated between geometric and spectroscopic parameters for P-Cl and S-Cl bonds

(Belskii, Naumov and Nuretdinov, 1974) and for Si-O, Ge-O, and
P-O bonds (Lazarev, Mirgorodskii, Ignatyev and Muldagaliev, 1975).
Since up-to-date and consistent geometric data have recently
become available for a relatively large series of simple
sulphones, a revision of the empirical relationships was timely
(Brunvoll and Hargittai, 1977b). Before presenting the results,
some comments are made concerning the data applied, the force
constant calculations and on the problem of averaging the $S=O$
stretching frequencies.

Geometric data. Structural parameters referring to vapour-phase
molecules were considered only. The relevant data are collected
in Table 26. Of the two investigations of SO_2F_2, the values for
the $S=O$ distance and O...O distance were taken from the electron
diffraction and microwave studies, respectively, and the $O=S=O$
bond angle was thus calculated. Although the physical meaning of
the internuclear distance parameters determined by the two techniques is different, the sulphone molecules under consideration
are relatively rigid systems and accordingly the electron diffraction and microwave spectroscopic distance parameters are
thought to be little different as a consequence of intramolecular
motion. Assumed parameters including bond angles $O=S=O$ calculated
using assumed O...O distance were not included in the determination of the relationships. The data on the ring molecules were
not utilized in some of the calculations.

Vibrational frequencies. The $S=O$ symmetric and antisymmetric stretching and SO_2 bending frequencies used in the calculations are listed in Table 40. They are similar to those used by Gillespie and Robinson (1963). They refer to liquid-phase infrared measurements unless otherwise stated. Although vapour-phase data would have been preferred, they are too scarce. It is difficult to predict the changes in frequencies corresponding to the transition from the vapour phase to the pure liquid or a dissolved state. It has been noted that stretching frequencies in liquids are often displaced to lower frequencies and bending frequencies to higher frequencies as compared with the vapour-phase data. Infrared gaseous and liquid data are quoted below from a study by Spoliti, Chackalackal and Stafford (1967) showing that no trend can be indeed established with certainty, at least for the stretching frequencies

		gas	liquid (condensed)	difference
CH_3SO_2F	$\nu_{as}(S=O)$, cm^{-1}	1451	1443	8
	$\nu_s(S=O)$, cm^{-1}	1225	1233	-8
	$\delta(SO_2)$, cm^{-1}	534	543	-10
CH_3SO_2Cl		1404	1386	18
		1192	1185	7
		543	548	-5
$CH_3SO_2CH_3$		1355	1343	12
		1162	1165	-3
		496	510	-14

The changes are relatively small, and significantly, the data originate from concurrent experiments. Referring to the dimethyl

142

Table 40

Experimental symmetric and antisymmetric stretching and bending frequencies of SO_2 groups in sulphone molecules. Calculated averages of the stretching frequencies (all in cm^{-1})

Compound	experimental data[*]			averages[**]			
	ν_s	ν_{as}	$\delta(SO_2)$	(1)	(2)	(3)	(4)
SO_2F_2	1269	1502	544[a]	1387	1394	1390	1386
SO_2Cl_2	1182	1414	560[b]	1295	1304	1303	1298
SO_2FBr	1205	1437	564[c]	1319	1327	1326	1321
SO_2FCl	1228	1455	480[a]	1343	1348	1346	1342
CH_3OSO_2F	1235	1465[d]			1357	1355	1350
CH_3SO_2F	1223	1415	531[e]	1320	1327	1322	1319
$C_6H_5SO_2Cl$	1186	1377	572[f]	1279	1289	1285	1282
CH_3OSO_2Cl	1192	1404	588[g]	1294	1304	1302	1298
$(CH_3)_2NSO_2Cl$	1182	1395	572[h]	1286	1295	1293	1289
CH_3SO_2Cl	1173	1378	533[i]	1273	1281	1280	1276
SO_2	1151	1362	518[j]	1253	1261	1261	1257
$(CH_3)_2NSO_2N(CH_3)_2$	1150	1335	526[h]	1240	1248	1246	1243
$(CH_3)_2SO_2$	1165	1343	510[k]	1252	1260	1257	1254
$(C_6H_5)_2SO_2$	1149	1305	580[l]	1223	1234	1229	1227
$(CH_2)_2SO_2$	1159	1309	446[m]	1237	1242	1236	1234
$(CH_2=CH)_2SO_2$	1124	1320	550[n]	1212	1221	1221	1217
$(CH_2)_4SO_2$	1147	1301	567[o]	1220	1230	1226	1224

[*] Liquid-phase infrared data unless indicated otherwise
[**] (1) and (2) Corresponding to expression (1) in text and using the f_r values of Columns (1) and (2) in Table 41, respectively (3) and (4) Corresponding to expressions (3)

Table 40 (continued)

and (4) in text, respectively

(a), (c), (m) Raman; (h) Solution; (j) Gaseous phase

(a) Birchall and Gillespie (1966)

(b) Raley, Wollrab and Lovejoy (1973)

(c) Reed and Lovejoy (1968)

(d) Dudley, Cady and Eggers (1956)

(e) Geiseler and Nagel (1973)

(f) Uno, Machida and Hanai (1968)

(g) Nagel, Stark, Fruwert and Geiseler (1976)

(h) Török, Páldi, Dobos and Fogarasi (1970)

(i) Cyvin, Dobos, Hargittai, Hargittai and Augdahl (1973)

(j) Shelton, Nielsen and Fletcher (1953)

(k) Spoliti, Chackalackal and Stafford (1967)

(l) Nagel, Steiger, Fruwert and Geiseler (1975)

(m) Kuz'yants and Aleksanyan (1972)

(n) Bulcke (1975)

(o) Katon and Feairheller (1965)

sulphone data, the above values are confronted below with some
which refer to solid-state infrared and solution Raman spectroscopy (Uno, Machida and Hanai, 1971):

	infrared	Raman
$\nu_{as}(S=O)$, cm^{-1}	1307	1289
$\nu_{s}(S=O)$, cm^{-1}	1143	1138
$\delta(SO_2)$, cm^{-1}	504	502

A final example, for gas-liquid comparison with considerable
differences is quoted from the infrared study of Bulcke (1975)
on divinyl sulphone

	gas	liquid	difference
$\nu_{as}(S{=}O)$, cm^{-1}	1352	1310	-42
$\nu_{s}(S{=}O)$, cm^{-1}	1150	1124	-26

Thus we may conclude that the next step for a future refinement of the relationships between geometric and vibrational parameters may necessitate a consistent set of vapour-phase vibrational spectroscopic data.

The S=O stretching force constants. Although the main purpose of establishing the empirical relationships was to predict geometric parameters from frequencies and vice versa, force constants were also calculated to provide a consistent set for them in the whole series. Even if all vibrational frequencies are known for a molecule, approximations have to be introduced, as a rule, to determine a force field. Thus the valence force constants are, to a certain extent, dependent on the approximations used. For several of the molecules discussed here, not even a complete set of experimental frequencies is available. Accordingly, further simplifications were necessary. The SO_2 group was looked upon as a separate entity, for which the well known expressions for a bent XY_2 model were applied (Cyvin, 1968).

As regards the approximations, two approaches were used. Approximation (1): G-matrix elements corresponding to kinematic coupling between the SO_2 group and the rest of the molecule were ignored; All force constants between the SO_2 group and the rest of the molecule were equal to zero; and The force constant for the SO_2 stretching-bending interaction was ignored.

Approximation (2): Same as approximation (1) and the G-matrix element connected with kinematic coupling between symmetric S=O stretching and O=S=O bending was also ignored.

The force constants calculated as outlined above are given in Table 41. The effects of the approximations used have been examined in detail (Brunvoll and Hargittai, 1977b).

On averaging the S=O stretching frequencies. Gillespie and Robinson (1963) used the average value of the antisymmetric and symmetric stretching frequencies in determining various correlation relationships. Szmant (1971) argued, however, that the two kinds of stretching frequencies are influenced to different extent by changes in the substituents on the sulphone group. Accordingly, the correlation of the antisymmetric and symmetric stretching frequencies and the influence of ways of averaging them were examined.

If the S=O group is considered to be a diatomic molecule, for this model

$$\lambda = (\frac{1}{m_S} + \frac{1}{m_O}) \; f_r \; , \tag{1}$$

where $\lambda = 4 \pi^2 c^2 \nu^2$, m_S and m_O are the sulphur and oxygen atomic masses. For the f_r force constant those found for the SO_2 group can be used. The $\nu(S=O)$ values given in Column (a) and (b) of Table 40 were calculated using expression (1) and force constants of Columns (1) and (2) of Table 41, respectively. Using approximation (2) for expressing the force constant f_r and utilizing that $m_S \approx 2 \, m_O$, we have

Table 41

Calculated stretching force constants, $f_r(S=O)$ (mdyn$Å^{-1}$)

Compound	(1)[a]	(2)[b]	Compound	(1)[a]	(2)[b]
SO_2F_2	12.081	12.202	CH_3SO_2Cl	10.175	10.305
SO_2Cl_2	10.540	10.680	SO_2	9.856	9.980
SO_2FBr	10.927[c]	11.068[c]	$(CH_3)_2NSO_2N(CH_3)_2$	9.654[c]	9.782[c]
SO_2FCl	11.329[c]	11.421[c]	$(CH_3)_2SO_2$	9.850	9.967
CH_3OSO_2F		11.574	$(C_6H_5)_2SO_2$	9.393[c]	9.560[c]
CH_3SO_2F	10.942	11.063	$(CH_2)_2SO_2$	9.608	9.690
$C_6H_5SO_2Cl$	10.281	10.432	$(CH_2=CH)_2SO_2$	9.228[c]	9.376[c]
CH_3OSO_2Cl	10.520	10.683	$(CH_2)_4SO_2$	9.349[c]	9.507[c]
$(CH_3)_2NSO_2Cl$	10.380	10.530			

[a] As described in the text as Approximation (1)

[b] As described in the text as Approximation (2)

[c] Bond angles used in these calculations were partly based on assumptions

$$= \frac{1}{2} \left[3\left(\frac{\nu_s}{\cos^2 A + 1} + \frac{\nu_{as}}{\sin^2 A + 1} \right) \right]^{1/2} \tag{2}$$

which gives then identical results with those presented in Column (2) of Table 41.

If SO_2 is considered and the average of λ_s and λ_{as} is taken for this triatomic molecule, ignoring $G_{12}(A_1)$ and f_{rr} we get

$$\nu = [\tfrac{1}{2}(\nu_s^2 + \nu_{as}^2)]^{1/2}, \qquad (3)$$

and the corresponding values are listed in Column (3) (Table 40). Finally Column (4) (Table 40) contains the values obtained by simple averaging

$$\nu = \tfrac{1}{2}(\nu_s + \nu_{as}) \qquad (4)$$

The differences between corresponding values in Columns (1) and (2) of Table 40 are between 5 and 11 cm^{-1}. The frequencies found by means of the average λ values, expression (3), given in Column (3) of Table 40 are all smaller than or equal to the frequencies of Column (2) of the same Table. The average frequencies, Column (4), all should be smaller than the corresponding values of Column (3) according to the comparison of eqs. (3) and (4). At the same time most of them are larger than the corresponding values of Column (1). Exceptions are SO_2F_2 and CH_3SO_2F (1 cm^{-1} lower), and $(CH_2)_2SO_2$ (3 cm^{-1} lower). No values deviate more than 5 cm^{-1} from the corresponding frequencies of Column (1). Thus the simple approach of taking the average frequency does not give a much different result from the method of looking upon the SO_2 group as a separate entity (cf. approximation (1), the section on the force constants), and computing the frequencies from the corresponding stretching force constants, cf. eq. (1) of this section. The frequencies found from the average values of the λ's are within 6 cm^{-1} from the results obtained by means of the S=O stretching force constants when in addition to the above approximations, $G_{12}(A_1)$ was also ignored, cf. approximation (2), the section on the force constants.

Correlation relationships.   A standard least-squares procedure
(Kuo, 1965) was used to establish the correlation relationships
among the geometric and vibrational parameters. The geometric
data were used with weights inversely proportional with the
experimental errors reported in the original studies. The
vibrational data were given equal weights.

The coefficients determined are listed in Table 42[*] for some
of the relationships including the one between the S=O bond
distances and O=S=O bond angles. These relationships are believed
to be useful within the rather narrow ranges covered by the
experimental data. The linear relationships are certainly
preferred for extrapolation rather than the second degree expression. Unfortunately, the parameter $r(S=O)$ of SO_2Cl_2 was seen not
to be consistent with the other data and thus sulphuryl chloride
was not considered in establishing the relationships listed in
Table 42 (except that between the symmetric and antisymmetric

---

[*] The data on sulphonyl chloride isocyanate have not been considered since its molecular geometry was not yet available at
the time of establishing these relationships. In fact the latter
were utilized as $r(S=O)$ was estimated on the basis of the
experimental stretching frequencies (ν_{as} 1412 cm^{-1} and
ν_s 1182 cm^{-1}; Kanesaka and Kawai, 1970) for the trial structure in the electron diffraction structure analysis of this
compound (Brunvoll, Hargittai and Seip, 1977).

frequencies). Inversely, 1.420 Å can be predicted for r(S=O) in
SO_2Cl_2[*]. Since the value of r(S=O) was estimated for SO_2FCl in its
microwave spectroscopic study by Holt and Gerry (1971) it was
interesting to observe that the present correlation relationship
provides firm support for their estimate[**]. On the other hand,
1.413 Å can be estimated for r_g(S=O) of SO_2FBr on the basis of the
stretching frequency vs. bond length relationship. The data on
$(CH_2)_2SO_2$ were also left out of most of our final calculations.
Their poorer agreement with the relationships for the other
parameters may be connected with different bonding peculiarities
in this rather strained system.

Some of the relationships established are given in Table 42
and shown in Figures 33, and 37-39.

---

[*] Comparing the earlier (Hargittai, 1968) and further refined
(Hargittai, 1969) values of r_g(S=O), it is suspected that this
parameter was strongly influenced by the empirical background
used (the change may be up to 0.01 Å). A reinvestigation is
warranted in any case.

[**] The value of 1.408 Å was estimated as to be between 1.405 Å
as given for SO_2F_2 in the microwave paper (Lide, Mann and
Fristrom, 1957) and 1.409 Å as given for SO_2Cl_2 in the earlier
electron diffraction paper (Hargittai, 1968). By fortunate
coincidence, the value of 1.408 Å is still between the electron
diffraction value of 1.398 Å for SO_2F_2 (Table 3) and the one
(1.420 Å) estimated now for SO_2Cl_2 .

Table 42

The constants A_o, A_1, and A_2 for $y = A_o + A_1x + A_2x^2$ found by the least-squares method. 2A is the O=S=O bond angle, R is the S=O bond length

y	x	A_o	A_1	A_2	standard deviation
2A	R	−3040.52	4632.13	−1693.57	0.56
2A	R	358.373	−166.525		0.61
lg 2A	lg R	2.37969	−1.92899		0.62(a)
ν_s	R	79031.6	−106631	36487.8	9.8
ν_s	R	5404.21	−2962.62		12.5
lg ν_s	lg R	3.61213	−3.50857		11.9(a)
ν_{as}	R	42738.9	−53962.3	17500.0	14.0
ν_{as}	R	7426.24	−4241.64		14.5
lg ν_{as}	lg R	3.80090	−4.29674		14.2(a)
ν_s	ν_{as}	3888.88	−4.48191	0.00182282	11.2
ν_s	ν_{as}	375.297	0.583985		13.0
lg ν_s	lg ν_{as}	0.947270	0.676814		13.1(a)
$f_r(b)$	R	1369.14	−1854.11	631.889	0.132
$f_r(b)$	R	95.1364	−59.5477		0.193
lg $f_r(b)$	lg R	6.88854	−69.1779		0.131(a)
$f_r(c)$	R	1200.76	−1615.04	547.123	0.158
$f_r(c)$	R	97.1161	−60.8273		0.196
lg $f_r(c)$	lg R	2.25034	−8.01743		0.176(a)

Table 42 (continued)

y	x	A_o	A_1	A_2	standard deviation
ν (d)	R	61159.2	−80683.0	27130.3	9.8
ν (d)	R	6432.34	−3614.05		11.3
lg ν (d)	lg R	3.71379	−3.94097		10.7

(a) This standard deviation does not correspond to y = lg z but to z itself; (b) and (c) These force constants correspond to Columns (1) and (2) in Table 41, respectively; (d) These frequencies correspond to Column (4) in Table 40

It is interesting to note that the relationship obtained directly from the data on 2A and R is consistent with similar relationships between these two parameters as obtained through a third parameter. Thus, through the symmetric stretching frequencies: $2A = 342.993 - 155.395\ R$, the asymmetric stretching frequencies: $2A = 343.371 - 155.520\ R$, and the force constants, $f_r(1)$: $2A = 353.663 - 163.242\ R$.

It has already been mentioned that 1.42 Å can be estimated for the S=O bond length in sulphuryl chloride on the basis of the experimental frequencies. This is illustrated in Figure 34. This Figure also shows that the S=O bond length of trichloromethyl sulphonyl chloride can be estimated on the basis of the frequencies to be also about 1.42 Å versus the reported 1.45±0.01 Å. Thus the

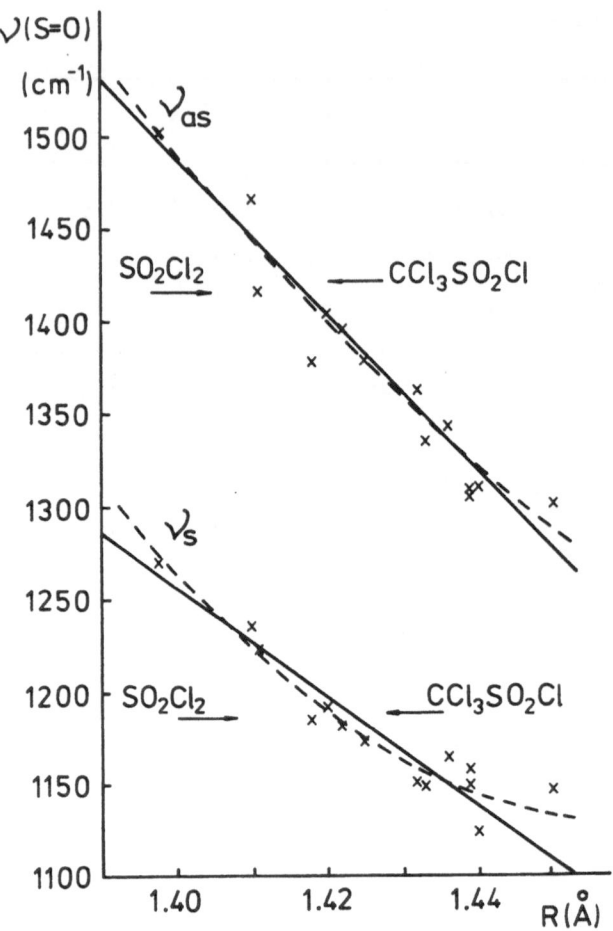

Figure 37

The symmetric and antisymmetric S=O stretching frequencies as
functions of the S=O bond length (R)
+ experimental values
—— linear equations (Table 42)
--- second degree equations (Table 42)
The S=O stretching frequencies of SO_2Cl_2 and CCl_3SO_2Cl are
indicated by arrows

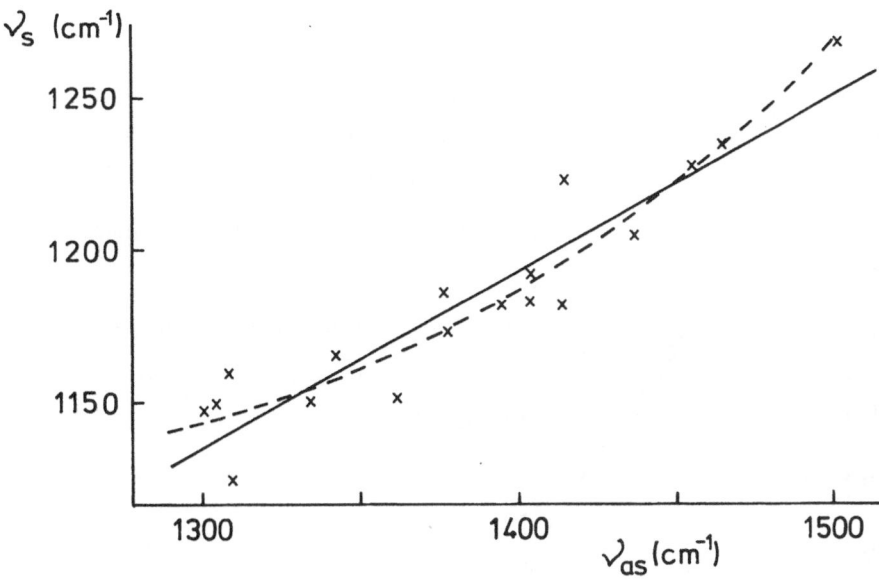

Figure 38

The symmetric S=O stretching frequency as a function of the antisymmetric S=O stretching frequency

+ experimental values

—— linear equation (Table 42)

--- second degree equation (Table 42)

bond lengths r(S=O) are expected to be the same in SO_2Cl_2 and CCl_3SO_2Cl indicating that substitution of Cl for the CCl_3 group has no appreciable effect. Accordingly, the S-Cl bond is expected to be of the same length as that in SO_2Cl_2 (2.01 Å). It has

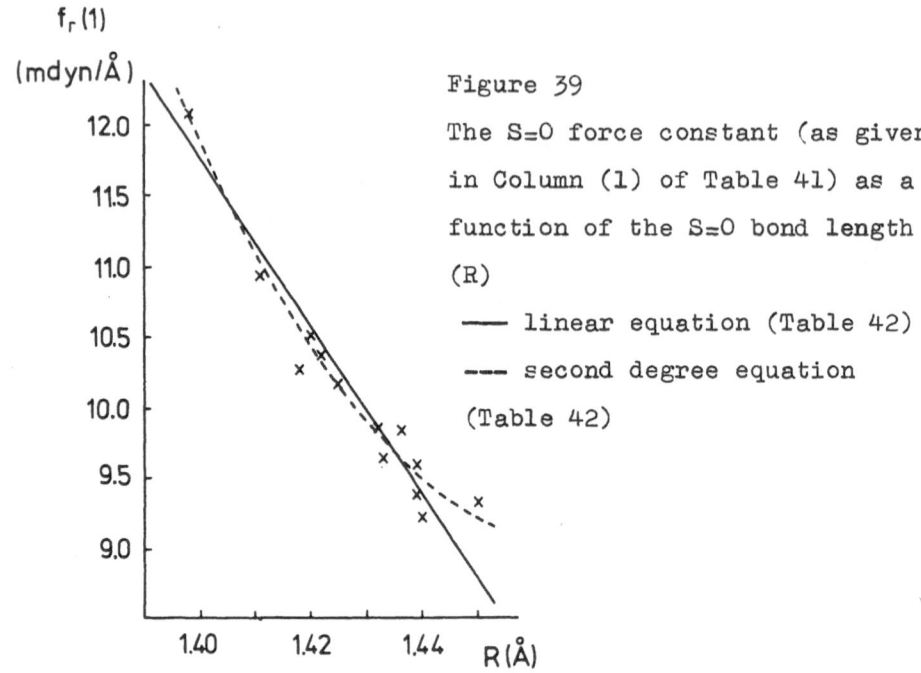

Figure 39

The S=O force constant (as given in Column (1) of Table 41) as a function of the S=O bond length (R)

——— linear equation (Table 42)

--- second degree equation (Table 42)

already been noted (Vajda and Hargittai, 1976) that the substitution of a chlorine atom by a trichloromethyl group has no apparent influence on the length of the Ge-Cl bonds as they are compared in $GeCl_4$ (2.113 Å; Morino, Nakamura and Iijima, 1960) and in CCl_3GeCl_3 (2.113 Å; Vajda, Hargittai, Maltsev and Nefedov, 1974). On the other hand methyl substitutions cause considerable lengthening of the Ge-Cl bonds in the methyl derivatives of germanium tetrachloride (cf. Vajda and Hargittai, 1976).

Our final note concerns the molecular geometry of sulphuryl fluoride. We refer to the differences in the S=O bond lengths

and O=S=O bond angles as they were determined by microwave
spectroscopy and electron diffraction and cited in our Table 3.
For reasons discussed in the section on the "Experimental
determination of the molecular geometries", the electron diffraction S=O bond length and the microwave spectroscopic O...O
distance may be preferred. Combining[*] the two, the O=S=O bond
angle is obtained to be 125.1^{o}. All this is consistent with
the correlations between the geometric and vibrational
parameters: The relationships between the S=O stretching frequencies and O=S=O bond angles are depicted in Figure 40 with
the standard deviations also indicated. The points marked as ED
and MW refer to the electron diffraction and microwave spectroscopic results and the vibrational frequencies of SO_2F_2 (cf.
Table 40). Both points are considerably off the curves representing the empirical relationships. However, when the O=S=O
bond angle calculated from the electron diffraction S=O distance
and microwave spectroscopic O...O distance is used the agreement
becomes strikingly good. This is marked by the circled cross.
Also, using the same curves, the bond angle O=S=O of CCl_3SO_2Cl
can be predicted to be between 122 and 123° versus the reported
111°.

---

[*] This is correct only, of course, to the extent to which the
differences in the physical meaning of the two kinds of
parameters may be ignored.

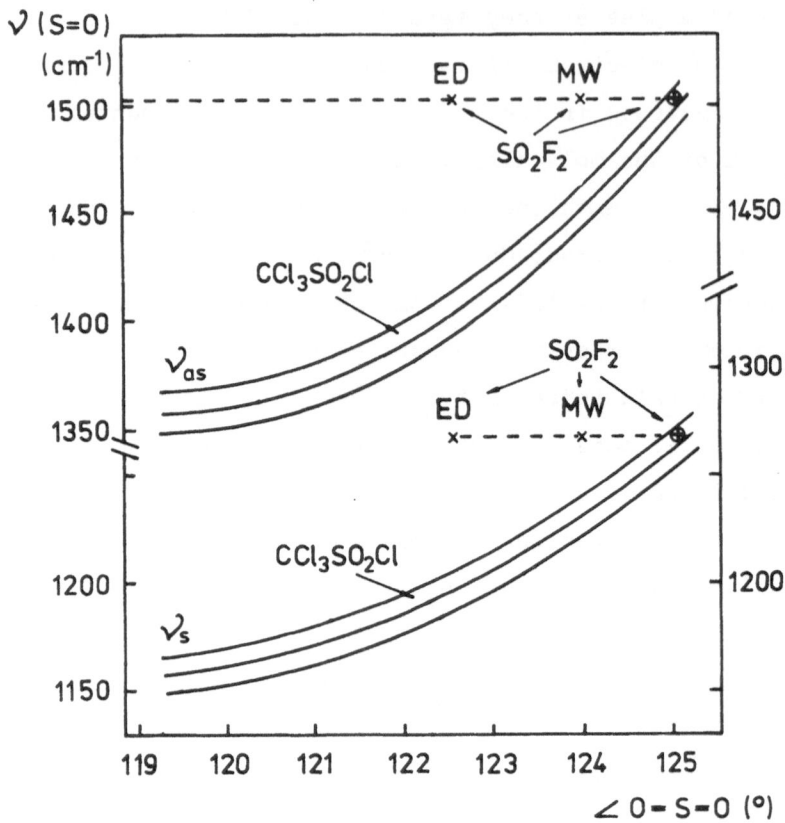

Figure 40

The symmetric and antisymmetric S=O stretching frequencies as
functions of the O=S=O bond angle (with the standard deviations).
The arrows indicate the S=O stretching frequencies of CCl_3SO_2Cl.
The dashed lines indicate the S=O stretching frequencies of
SO_2F_2 . For the explanation of the markings labelled by ED and
MW and the circled cross, see the text

REFERENCES

Abbar, Ch. (1963): Comptes Rendus Acad. Sci. Paris, 257, 2022.

Abbar, Ch. (1965): Comptes Rendus Acad. Sci. Paris, 261, 365.

Abbar, Ch., Journel, G. and Moise, A. (1965): Comptes Rendus
Acad. Sci. Paris, 261, 5047.

Airey, W., Glidewell, C., Rankin, D.W.H., Robiette, A.G.,
Sheldrick, G.M. and Cruickshank, D.W.J. (1970): Trans. Faraday
Soc., 66, 551.

Airey, W., Glidewell, C., Robiette, A.G. and Sheldrick, G.M.
(1971a): J. Mol. Structure, 8, 413.

Airey, W., Glidewell, C., Robiette, A.G. and Sheldrick, G.M.
(1971b): J. Mol. Structure, 8, 435.

Airey, W., Glidewell, C., Robiette, A.G., Sheldrick, G.M. and
Freeman, J. (1971): J. Mol. Structure, 8, 423.

Alekseev, N.V. (1967): Zh. Strukt. Khimii, 8, 532.

Alléaume, M. and Decap, J. (1965a): Acta Crystallogr., 18, 731.

Alléaume, M. and Decap, J. (1965b): Acta Crystallogr., 19, 934.

Alléaume, M. and Decap, J. (1968): Acta Crystallogr., B24, 214.

Almenningen, A., Anfinsen, A. and Haaland, A. (1970): Acta Chem.
Scand., 24, 43.

Anderson, D.W.W., Rankin, D.W.H. and Robertson, A. (1972):
J. Mol. Structure, 14, 385.

Arnold, D.E.L. and Rankin, D.W.H. (1972/1973): J. Fluorine
Chem., 2, 405.

Banister, A.J., Moore, L.F. and Padley, J.S. (1968):
Structural Studies on Sulphur Species. Inorganic Sulphur
Chemistry (ed.: Nickless, G.). Elsevier, Amsterdam.

Bartell, L.S. (1972): Electron Diffraction by Gases. Physical Methods in Organic Chemistry (ed.: Weissberger, A. and Rossiter, B.W.). Interscience, New York.

Bartell, L.S. and Bonham, R.A. (1960): J. Chem. Phys., 32, 824.

Bartell, L.S. and Brockway, L.O. (1960): J. Chem. Phys., 32, 512.

Bartell, L.S. and Hirst, L.C. (1959): J. Chem. Phys., 31, 449.

Bartlett, N., Wechsberg, M., Jones, G.R. and Burbank, R.D. (1972): Inorg. Chem., 11, 1124.

Bauer, S.H. (1970): Diffraction of Electrons by Gases. Physical Chemistry, Vol. 4 (ed.: Henderson, D.). Academic Press, New York.

Beagley, B., Clark, A.H. and Hewitt, T.G. (1968): J. Chem. Soc., 658.

Beagley, B. and Conrad, A.R. (1970): Trans. Faraday Soc., 66, 2740.

Beagley, B. and Hewitt, T.G. (1968): Trans. Faraday Soc., 64, 2561.

Beall, R., Herdklotz, J. and Sass, R.L. (1970): Acta Crystallogr., B26, 1633.

Belskii, V.E., Naumov, V.A. and Nuretdinov, I.A. (1974): Dokl. Akad. Nauk S.S.S.R., 215, 355.

Bezzubov, V.M. and Naumov, V.A. (1976): Zh. Strukt. Khimii, 17, 98.

Birchall, T. and Gillespie, R.J. (1966): Spectrochimica Acta, 22, 681.

Boggs, J.E. (1975): private communication.

Bouchy, A. and Roussy, G. (1973): Comptes Rendus Acad. Sci. Paris, 277, 143.

Boys, S.F. (1966): in Quantum Theory of Atoms, Molecules and the Solid State (ed.: Lövdin, P.O.). Academic Press, New York.

Brittain, A.H., Smith, J.E., Lee, P.L., Cohn, K. and Schwendeman, R.H. (1971): J. Amer. Chem. Soc., 93, 6772.

Brooks, W.V.F. and Cyvin, S.J. (1962): Acta Chem. Scand., 16, 820.

Brooks, W.V.F., Cyvin, S.J. and Kvande, P.C. (1965): J. Phys. Chem., 69, 1489.

Brunvoll, J. and Hargittai, I. (1976): J. Mol. Structure, 30, 361.

Brunvoll, J. and Hargittai, I. (1977a): Acta Chim. (Budapest), in the press.

Brunvoll, J. and Hargittai, I. (1977b): Acta Chim. (Budapest), submitted.

Brunvoll, J., Hargittai, I. and Seip, R. (1977): to be published.

Bulcke, P. (1975): Diplomarbeit, KMU Leipzig.

Bürger, H., Burczyk, K., Blascheffe, A. and Safari, H. (1971): Spectrochim. Acta, 27A, 1073.

Bürgi, H.B., Stedman, D. and Bartell, L.S. (1971): J. Mol. Structure, 10, 31.

Carlos, J.L. and Bauer, S.H. (1974): J. Chem. Soc. Faraday Trans. II, 70, 171.

Carlos, J.L., Karl, R.R. and Bauer, S.H. (1974): J. Chem. Soc. Faraday Trans. II, 70, 177.

Caron, A., Palenik, G.J., Goldish, E. and Donohue, J. (1964):
Acta Crystallogr., 17, 102.

Casado, J., Nygaard, L. and Sørensen, G.O. (1971): J. Mol.
Structure, 8, 211.

Cazzoli, G., Lister, D.J. and Favero, P.J. (1972): J. Mol.
Spectroscopy, 42, 286.

Churchill, M.R., Deboer, B.G. and Kalra, K.L. (1972): J.C.S.
Chem. Commun., 981.

Clark, A.H. and Beagley, B. (1971): Trans. Faraday Soc., 67,
2216.

Clark, A.H. and Haaland, A. (1970): Acta Chem. Scand., 24,
3024.

Clever, H.L. and Westrum, E.F. (1970): J. Phys. Chem., 74,
1309.

Codding, E.G., Jones, C.E. and Schwendeman, R.H. (1974):
Inorg. Chem., 13, 178.

Cohen, E.R. and Taylor, B.N. (1973): Phys. Chem. Ref. Data,
2, 663.

Corosine, M., Crasnier, M.-C., Labarre, M.-C., Labarre, J.-F.
and Leibovici, C. (1973): Chem. Phys. Lett., 20, 111.

Cotton, F.A. and Stokely, P.F. (1970): J. Amer. Chem. Soc.,
92, 294.

Csákvári, B., Wagner, Zs., Gömöry, P., Hargittai, I.,
Rozsondai, B. and Mijlhoff, F.C. (1976): Acta Chim. (Budapest),
90, 149.

Csákvári, B., Wagner, Zs., Gömöry, P., Mijlhoff, F.C.,
Rozsondai, B. and Hargittai, I. (1976): J. Organometal.
Chem., 107, 287.

Cunningham, G.L. Boyd, A.W., Myers, R.J. and Gwinn, W.D.
(1951): J. Chem. Phys., 19, 676.

Cyvin, B.N. (1975): private communication.

Cyvin, B.N. and Cyvin, S.J. (1972): Acta Chem. Scand., 26,
1284.

Cyvin, S.J. (1968): Molecular Vibrations and Mean Square
Amplitudes. Universitetsforlaget, Oslo and Elsevier,
Amsterdam.

Cyvin, S.J. (1973): private communication.

Cyvin, S.J., Dobos, S., Hargittai, I., Hargittai, M. and
Augdahl, E. (1973): J. Mol. Structure, 18, 203.

Cyvin, S.J. and Hargittai, I. (1968): unpublished calculations as cited by Hargittai (1974a).

Cyvin, S.J. and Hargittai, I. (1969): Acta Chim. (Budapest),
61, 159.

Domenicano, A., Mazzeo, P. and Vaciago, A. (1976):
Tetrahedron Letters, 1029.

Domenicano, A., Vaciago, A. and Coulson, C.A. (1975): Acta
Crystallogr., B31, 221; 1630.

Dorney, A.J., Hoy, A.R. and Mills, I.M. (1973): J. Mol.
Spectroscopy, 45, 253.

Dousmanis, G.C., Sanders, T.M., Townes, C.H. and Zeiger, H.
J. (1953): J. Chem. Phys., 21, 1416.

Dubrulle, A. and Boucher, D. (1974): Comptes Rendus Acad. Sci. Paris 278 Serie B, 211.

Dudley, F.B., Cady, G.H. and Eggers, D.F. (1956): J. Amer. Chem. Soc., 78, 290.

Dupont, P.L. and Dideberg, O. (1972): Acta Crystallogr., B28, 2340.

Eisma, S.W., Altona, C., Geise, H.J., Mijlhoff, F.C. and Renes, G.H. (1974): J. Mol. Structure, 20, 251.

Exner, O., Dembech, P. and Vivarelli, P. (1971): J. Chem. Soc. A, 620.

Feder, W., Dreizler, H., Rudolph, H.D. and Typke, V. (1969): Z. Naturforsch., 24a, 266.

Forti, P., Damiani, D. and Favero, P.G. (1973): J. Amer. Chem. Soc., 95, 756.

Gebhardt, O. (1971): Dissertation, Trondheim.

Geiseler, G. and Nagel, B. (1973): J. Mol. Structure, 16, 79.

Gergő, É., Hargittai, I. and Schultz, Gy. (1976): J. Organometal. Chem., 112, 29.

Gerry, M.C.L., Thompson, J.C. and Sugden, T.M. (1966): Nature, 211, 846.

Gillespie, R.J. (1972): Molecular Geometry. Van Nostrand Reinhold, London.

Gillespie, R.J. and Nyholm, R.S. (1957): Quart. Rev. (London), 11, 339.

Gillespie, R.J. and Robinson, E.A. (1961): Can. J. Chem., 39, 2171.

Gillespie, R.J. and Robinson, E.A. (1963): Can. J. Chem., 41, 2074.

Glidewell, C., Rankin, D.W.H., Robiette, A.G. and Sheldrick, G.M.
(1969): J. Mol. Structure, 4, 215.

Glidewell, C., Rankin, D.W.H., Robiette, A.G. and Sheldrick, G.M.
(1970a): J. Chem. Soc. A, 318.

Glidewell, C., Rankin, D.W.H., Robiette, A.G. and Sheldrick, G.M.
(1970b): J. Mol. Structure, 6, 231.

Glidewell, C., Robiette, A.G. and Sheldrick, G.M. (1972): Chem.
Phys. Lett., 16, 526.

Gogoi, B.N. and Hargreaves, A. (1970): Acta Crystallogr., B26, 2132.

Gordy, W. and Cook, R.L. (1970): Microwave Molecular Spectra.
Interscience, New York.

Gropen, O. and Seip, H.M. (1971): Chem. Phys. Lett., 11, 445.

Guin, H.W. (1969): Dissertation. Austin, Texas. Private
communication from Simonsen, S.H. (1975).

Gundersen, G. (1976): J. Mol. Structure, 33, 79.

Gundersen, G. and Vahrenkamp, H. (1976): J. Mol. Structure,
33, 97.

Hagen, K., Coussens, V.R. and Hedberg, K. (1975): unpublished
results, private communication.

Ham, N.S., Hambly, A.N. and Laby, R.H. (1960): Aust. J. Chem.,
13, 443.

Hamilton, W.C. (1964): Statistics in Physical Science. The
Ronald Press Co., New York.

Hamilton, W.C. (1965): Acta Crystallogr., 18, 502.

Hargittai, I. (1968): Acta Chim. (Budapest), 57, 403.

Hargittai, I. (1969): Acta Chim. (Budapest), 60, 231.

Hargittai, I. (1973): Természet Világa, 104, 78.

Hargittai, I. (1974a): The Electron Diffraction Interatomic Distance (In Hungarian). A kémia ujabb eredményei ʿAdvances in Chemistry, ed.: Csákvári, B.), Vol. 21. Akadémiai Kiadó, Budapest.

Hargittai, I. (1974b): The Geometries of Tetrahedral and Related Molecules and the V.S.E.P.R. Model. Second European Crystallographic Meeting, Keszthely, Collected Abstracts, pp. 441–443.

Hargittai, I. and Baranyi, A. (1977): Acta Chim. (Budapest), in the press.

Hargittai, I. and Brunvoll, J. (1976): Acta Chem. Scand. A, 30, 634.

Hargittai, I. and Cyvin, S.J. (1969): Acta Chim. (Budapest), 61, 51.

Hargittai, I. and Hargittai, M. (1973a): J. Mol. Structure, 15, 399.

Hargittai, I. and Hargittai, M. (1973b): Acta Chim. (Budapest), 75, 129.

Hargittai, I. and Mijlhoff, F.C. (1973): J. Mol. Structure, 16, 69.

Hargittai, I. and Paul, I.C. (1977): Structural Chemistry of the Cyanates and their Thio Derivatives. The Chemistry of the Cyanates and their Thio Derivatives (ed.: Patai, S.). Wiley, London.

Hargittai, I., Rozsondai, B., Nagel, B., Bulcke, P., Robinet, G. and Labarre, J.-F. (1977): to be published.

Hargittai, I., Schultz, Gy. and Kolonits, M. (1977): J. Chem.
Soc. Dalton Trans., in the press.

Hargittai, I., Seip, R., Nair, K.P.R., Britt, Ch.O., Boggs,
J.E. and Cyvin, B.N. (1977): J. Mol. Structure, in the press.

Hargittai, I. and Vajda, E. (1975): unpublished calculations.

Hargittai, I., Vajda, E. and Szőke, A. (1973): J. Mol.
Structure, 18, 381.

Hargittai, I. and Vilkov, L.V. (1970): Acta Chim. (Budapest),
63, 143.

Hargittai, M. and Hargittai, I. (1973c): J. Chem. Phys., 59,
2513.

Hargittai, M. and Hargittai, I. (1974): J. Mol. Structure, 20,
283.

Hargittai, M. and Hargittai, I. (1977): The Molecular
Geometries of Coordination Compounds in the Vapour Phase.
Elsevier, Amsterdam and Akadémiai Kiadó, Budapest.

Harlow, R.L., Sammes, M.P. and Simonsen, S.H. (1974): Acta
Crystallogr., B30, 2903.

Harlow, R.L., Simonsen, S.H., Pfluger, C.E. and Sames, M.P.
(1974): Acta Crystallogr., B30, 2264.

Hedberg, K. (1955): J. Amer. Chem. Soc., 77, 6491.

Hedberg, K. and Iwasaki, M. (1962): J. Chem. Phys., 36, 589.

Helm, D. v.d., Childs, J.D. and Christian, S.D. (1969):
J. Chem. Soc. Chem. Commun., 887.

Hencher, J.L. and Bauer, S.H. (1973): Can. J. Chem., 51, 2047.

Hilderbrandt, R.L., Andreassen, A.L. and Bauer, S.H. (1970):
J. Phys. Chem., 74, 1586.

Hilderbrandt, R.L. and Bauer, S.H. (1969): J. Mol. Structure, 3, 325.

Hjortaas, K.E. (1967): Acta Chem. Scand., 21, 1381.

Hocking, W.H. and Gerry, M.C.L. (1975): J. Mol. Spectroscopy, 58, 250.

Hocking, W.H. and Gerry, M.C.L. (1976): J. Mol. Spectroscopy, 59, 338.

Hoffmann, R., Fujimoto, H., Swenson, J.R. and Wan, Ch.-Ch. (1973): J. Amer. Chem. Soc., 95, 7644.

Holt, C.W. and Gerry, M.C.L. (1971): Chem. Phys. Lett., 9, 621.

Jacob, E.J. and Lide, D.R. (1971): J. Chem. Phys., 54, 4591.

Jandacek, R.J. (1968): Dissertation. Austin, Texas. Private communication from Simonsen, S.H. (1975).

Jandal, P., Seip, H.M. and Torgrimsen, T. (1976): J. Mol. Structure, 32, 369.

Jenkins, D.R., Kewley, R. and Sugden, T.M. (1962): Trans. Faraday Soc., 58, 1284.

Jones, L.H., Shoolery, J.N., Shulman, R.G. and Yost, D.M. (1950): J. Chem. Phys., 18, 990.

Jordan, T., Smith, H.W., Lohr, L.L. and Lipscomb, W.N. (1963): J. Amer. Chem. Soc., 85, 846.

Kaldor, A. and Maki, A.G. (1973): J. Mol. Structure, 15, 123.

Kálmán, A., Duffin, B. and Kucsman, Á. (1971): Acta Crystallogr., B27, 586.

Kanesaka, I. and Kawai, K. (1970): Bull. Chem. Soc. Japan, 43, 3298.

Karle, I.L. (1973): Z. Kristallogr., 138, 184.

Karle, J. (1973): Electron Diffraction. Determination of Organic Structures by Physical Methods, Vol. 5. Academic Press, New York.

Katon, J.E. and Feairheller, W.R. (1965): Spectrochimica Acta, $\underline{21}$, 199.

Kiers, C. Th. and Vos, A. (1972): Rec. Trav. Chim., $\underline{91}$, 126.

Kimura, K., Katada, K. and Bauer, S.H. (1966): J. Amer. Chem. Soc., $\underline{88}$, 416.

Kimura, K. and Kubo, M. (1959): J. Chem. Phys., $\underline{30}$, 151.

Kirchhoff, W.H., Johnson, D.R. and Powell, F.X. (1973): J. Mol. Spectroscopy, $\underline{48}$, 157.

Klug, H.P. (1968): Acta Crystallogr., $\underline{B24}$, 792.

Kronfeld, L.R. and Sass, R.L. (1968): Acta Crystallogr., $\underline{B24}$, 981.

Kuchitsu, K. (1966): J. Chem. Phys., $\underline{44}$, 906.

Kuchitsu, K. (1968): J. Chem. Phys., $\underline{49}$, 4456.

Kuchitsu, K. and Cyvin, S.J. (1972): Representation and Experimental Determination of the Geometry of Free Molecules. Molecular Structures and Vibrations (ed.: Cyvin, S.J.). Elsevier, Amsterdam.

Kuchitsu, K., Guillory, J.P. and Bartell, L.S. (1968): J. Chem. Phys., $\underline{49}$, 2488.

Kuchitsu, K., Shibata, S., Yokozeki, A. and Matsumura, C. (1971): Inorg. Chem., $\underline{10}$, 2584.

Kuo, S.S. (1965): Numerical Methods and Computers. Addison-Wesley, Reading, Massachusetts, 1965.

Kuz'yants, G.M. and Aleksanyan, V.T. (1972): Zh. Strukt.
Khimii, 13, 617.

Langs, D.A., Silverton, J.V. and Bright, W.M. (1970): J. Chem.
Soc. Chem. Commun., 1653.

La Placa, J.J. and Ibers, J.A. (1966): Inorg. Chem., 5, 405.

Lazarev, A.N., Mirgorodskii, A.P., Ignatyev, I.C. and
Muldagaliev, Kh.Kh. (1975): Zh. Strukt. Khimii, 16, 578.

Lide, D.R. (1969): Survey of Progress in Chemistry 5. Academic
Press, New York.

Lide, D.R., Mann, D.E. and Fristrom, R.M. (1957): J. Chem.
Phys., 26, 734.

Lister, D.G., Tyler, J.K., Høg, J.H. and Larsen, N.W. (1974):
J. Mol. Structure, 23, 253.

Loghry, R.A. and Simonsen, S.H. (1975): private communication.

Lowenstein, M.Z. (1965): Diss. Abstr., 26, 2500.

Lucas, N.J.D. and Smith, J.G. (1972): J. Mol. Spectroscopy,
43, 327.

McDonald, W.S. and Cruickshank, D.W.J. (1967): Acta Crystallogr.,
22, 48.

Michel, F., Nery, H., Nosberger, P. and Roussy, G. (1976):
J. Mol. Structure, 30, 409.

Morino, Y., Kikuchi, Y., Saito, S. and Hirota, E. (1964):
J. Mol. Spectroscopy, 13, 95.

Morino, Y., Kuchitsu, K. and Moritani, T. (1969): Inorg.
Chem., 8, 867.

Morino, Y., Murata, Y., Ito, T. and Nakamura, J. (1962):
J. Phys. Soc. Japan, 17, BII, 37.

Morino, Y. Nakamura, Y. and Iijima, T. (1960): J. Chem. Phys., 32, 643.

Morino, Y. and Saito, S. (1966): J. Mol. Spectroscopy, 19, 435.

Moritani, T. Kuchitsu, K. and Morino, Y. (1971): Inorg. Chem., 10, 344.

Murdoch, J.D., Rankin, D.W.H. and Beagley, B. (1976): J. Mol. Structure, 31, 291.

Murray, J.T. Williams, Q. and Weatherly, T.L. (1972): Bull. Am. Phys. Soc., 17, 575A.

Nagel, B., Stark, J., Fruwert, J. and Geiseler, G. (1976): Spectrochimica Acta, 32A, 1297.

Nagel, B., Steiger, Th., Fruwert, J. and Geiseler, G. (1975): Spectrochimica Acta, 31A, 255.

Nakano, Y., Saito, S. and Morino, Y. (1970): Bull. Chem. Soc. Japan, 43, 368.

Naumov, V.A., Gulyaeva, N.A. and Pudovick, M.A. (1972): Dokl. Akad. Nauk S.S.S.R., 203, 590.

Naumov, V.A. and Shaidulin, S.A. (1976): Zh. Strukt. Khimii, 17, 304.

Naumov, V.A., Semashko, V.N. and Shaidulin, S.A. (1973): Zh. Strukt. Khimii, 14, 595.

Naumov, V.A., Semashko, V.N. and Shatrukov, L.F. (1973): Dokl. Akad. Nauk S.S.S.R., 209, 118.

Noordik, J.H. and Vos, A. (1967): Rec. Trav. Chim., 86, 156.

Nygaard, L., Bojesen, I., Pedersen, T. and Rastrup-Andersen, J. (1968): J. Mol. Structure, 2, 209.

Oberhammer, H. and Zeil, W. (1970): J. Mol. Structure, 6, 399.

O'Connel, A.M. and Maslen, E.N. (1967): Acta Crystallogr., 22, 134.

O'Connor, B.H. and Maslen, E.N. (1965): Acta Crystallogr., 18, 363.

Olie, K. and Mijlhoff, F.C. (1969): Acta Crystallogr., B25, 974.

Otake, M., Matsumura, C. and Morino, Y. (1968): J. Mol. Spectroscopy, 28, 316.

Pauling, L. (1960): The Nature of the Chemical Bond. Cornell University Press, Ithaca, New York.

Pedersen, T. (1969): J. Mol. Structure, 4, 59.

Perahia, D. and Pullman, A. (1973): Chem. Phys. Lett., 19, 73.

Pichai, R., Krishna Pillai, M.G. and Ramaswamy, K. (1967): Aust. J. Chem., 20, 1055.

Plato, V., Hartford, W.D. and Hedberg, K. (1970): J. Chem. Phys., 53, 3488.

Pople, J.A. and Beveridge, D.L. (1970): Approximate Molecular Orbital Theory. McGraw-Hill, New York.

Pulay, P.(1969): Mol. Phys., 17, 197.

Radnai, T., Kolonits, M., Gregory, D.C. and Hargittai, I. (1975): unpublished electron diffraction results cited by Rozsondai, Moore, Gregory and Hargittai (1977).

Raley, J.M., Wollrab, J.E. and Lovejoy, L.W. (1973): J. Mol. Spectroscopy, 48, 100.

Ralowski, W., Ljunggren, S. and Mjöberg, J. (1973): Acta Chem. Scand., 27, 3128.

Rankin, D.W.H. and Cyvin, S.J. (1972): J. Chem. Soc. Dalton Trans., 1277.

Ramaswamy, K. and Jayaraman, S. (1971): J. Mol. Structure, 10, 183.

Reed, P.R. and Lovejoy, R.W. (1968): Spectrochimica Acta, 24A, 1795.

Robinet, G., Crasnier, F., Labarre, J.-F. and Leibovici, C. (1972): Theoret. chim. Acta (Berl.), 25, 259.

Rozsondai, B., Moore, J.H., Gregory, D.C. and Hargittai, I. (1977): Acta Chim. (Budapest), in the press.

Sadova, N.I., Vilkov, L.V., Hargittai, I. and Brunvoll, J. (1976): J. Mol. Structure, 31, 131.

Saito, S. (1969): Bull. Chem. Soc. Japan, 42, 663.

Saito, S. (1970): Bull. Chem. Soc. Japan, 43, 368.

Saito, S. and Makino, F. (1972): Bull. Chem. Soc. Japan, 45, 92.

Samdal, S. and Seip, H.M. (1975): J. Mol. Structure, 28, 193.

Sands, D.E. (1963): Z. Kristallogr., 119, 245.

Sands, D.E. (1972): Acta Crystallogr., B28, 2463.

Sands, D.E. and Day, V.W. (1967): Z. Kristallogr., 124, 220.

Schäfer, L. (1976): Applied Spectroscopy, 30, 123.

Schmiedekamp, A., Skaarup, S., Pulay, P., Hargittai, I., Cruickshank, D.W.J. and Boggs, J.E. (1977): to be published.

Schomaker, V. and Stevenson, D.P. (1941): J. Amer. Chem. Soc., 63, 37.

Schultz, Gy. (1976): private communication.

Seip, H.M. (1973): Theory; Accuracy. Molecular Structure by Diffraction Methods. Specialist Periodical Reports (senior reporters: Sim, G.A. and Sutton, L.E.). The Chemical Society, London.

Seip, R., Schultz, Gy., Hargittai, I. and Szabó, Z.G. (1977): to be published.

Shelton, R.D., Nielsen, A.H. and Fletcher, W.H. (1953): J. Chem. Phys., 21, 2178.

Shibata, S. and Bartell, L.S. (1965): J. Chem. Phys., 42, 1147.

Shriver, D.F. (1970): Accounts of Chemical Research, 3, 231.

Sidgwick, N.V. and Powell, H.E. (1940): Proc. Royal Soc., 197, 153.

Sime, J.G. and Woodhouse, D.I. (1974): J. Cryst. Mol. Structure, 4, 269.

Shelton, R.D., Nielsen, A.H. and Fletcher, W.H. (1953): J. Chom. Phys., 21, 2178.

Spoliti, M., Chackalackal, S.M. and Stafford, F.E. (1967): J. Amer. Chem. Soc., 89, 1092.

Stevenson, D.P. and Schomaker, V. (1940): J. Amer. Chem. Soc., 62, 1913.

Sugden, T.M. and Kenney, C.N. (1965): Microwave Spectroscopy of Gases. Van Nostrand, London.

Sutton, L.E., editor (1965): Tables of Interatomic Distances and Configuration in Molecules and Ions, Spec. Publ. 18, The Chem. Soc., London.

Szmant, H.H. (1971): The Sulfur-Oxygen Bond. Sulfur in Organic and Inorganic Chemistry (ed.: Senning, A.), Vol. 1. Dekker, New York.

Tamagawa, K., Iijima, T. and Kimura, M. (1976): J. Mol. Structure, 30, 243.

Tokue, I., Fukuyama, T. and Kuchitsu, K. (1974): J. Mol. Structure, 23, 33.

Towns, R.L. and Simonsen, S.H. (1974): Crystal Structure Communications, 3, 373.

Towns, R.L. and Simonsen, S.H. (1975): Crystal Structure Communications,

Török, F., Páldi, E., Dobos, S. and Fogarasi, G. (1970): Acta Chim. (Budapest), 63, 417.

Trueblood, K.N. and Mayer, S.W. (1956): Acta Crystallogr., 9, 628.

Uno, T., Machida, K. and Hanai, K. (1968): Spectrochimica Acta, 24A, 1705.

Uno, T., Machida, K. and Hanai, K. (1971): Spectrochimica Acta, 27A, 107.

Vajda, E. and Hargittai, I. (1976): Acta Chim. (Budapest), 91, 185.

Vajda, E., Hargittai, I., Maltsev, A.K. and Nefedov, O.M. (1974): J. Mol. Structure, 23, 417.

Van Eijck, B.P., Korthoff, A.J. and Mijlhoff, F.C. (1975): J. Mol. Structure, 24, 222.

Van Wazer, J.R. and Absar, I. (1972): Advances in Chemistry, 110, 20.

Venkateswarlu, K. and Malathy Devi, V. (1965): Indian Pure Appl. Phys., 3, 195.

Vilkov, L.V. et al. (1974): The Theoretical Foundations of
Gas Electron Diffraction (in Russian). Moscow State University,
Moscow.

Vilkov, L.V. and Hargittai, I. (1967): Acta Chim. (Budapest),
52, 423.

Vilkov, L.V. and Khaikin, L.S. (1966): Dokl. Akad. Nauk
S.S.S.R., 168, 810.

Vilkov, L. and Khaikin, L.S. (1975): Stereochemistry of
Compounds Containing Bonds between Si, P, S, Cl and N or O.
Gas-Phase Electron Diffraction. Topics in Current Chemistry
(Managing Ed.: Boschke, F.), 53, 25. Springer-Verlag, Berlin,
Heidelberg, New York.

Vilkov, L.V., Khaikin, L.S. and Evdokimov, V.V. (1972): Zh.
Strukt. Khimii, 13, 7.

Vilkov, L.V., Penionzhkevich, N. P. , Brunvoll, J. and
Hargittai, I. (1977): to be published.

Vilkov, L.V. and Tarasenko, N.A. (1969): J.C.S. Chem. Commun.,
1176.

Vogt, L.H., Katz, J.L. and Wiblerley, S.E. (1965): Inorg.
Chem., 4, 231.

Wadhawan, V.K. (1976): Acta Crystallogr., B32, 397.

Wang, H.K. (1965): Acta Chem. Scand., 19, 879.

Wollrab, J.E. (1967): Rotational Spectra and Molecular
Structure. Academic Press, New York.

Yokozeki, A. and Bauer, S.H. (1975): The Geometric and
Dynamic Structures of Fluorocarbons and Related Compounds.

Gas-Phase Electron Diffraction. Topics in Current Chemistry
(Managing Ed.: Boschke, F.), 53, 71. Springer-Verlag, Berlin,
Heidelberg, New York.

Zaripov, N.M., Naumov, V.A. and Tuzova, L.L. (1974):
Phosphorus, 4, 179.